Methods in Enzymology

Volume 379
ENERGETICS OF BIOLOGICAL
MACROMOLECULES
Part D

METHODS IN ENZYMOLOGY

EDITORS-IN-CHIEF

John N. Abelson   Melvin I. Simon

DIVISION OF BIOLOGY
CALIFORNIA INSTITUTE OF TECHNOLOGY
PASADENA, CALIFORNIA

FOUNDING EDITORS

Sidney P. Colowick and Nathan O. Kaplan

Methods in Enzymology

Volume 379

Energetics of Biological Macromolecules

Part D

EDITED BY

Jo M. Holt

DEPARTMENT OF BIOCHEMISTRY AND MOLECULAR BIOPHYSICS
WASHINGTON UNIVERSITY SCHOOL OF MEDICINE
ST. LOUIS, MISSOURI

Michael L. Johnson

DEPARTMENT OF PHARMACOLOGY
UNIVERSITY OF VIRGINIA
CHARLOTTESVILLE, VIRGINIA

Gary K. Ackers

DEPARTMENT OF BIOCHEMISTRY AND MOLECULAR BIOPHYSICS
WASHINGTON UNIVERSITY SCHOOL OF MEDICINE
ST. LOUIS, MISSOURI

ELSEVIER
ACADEMIC
PRESS

AMSTERDAM • BOSTON • HEIDELBERG • LONDON
NEW YORK • OXFORD • PARIS • SAN DIEGO
SAN FRANCISCO • SINGAPORE • SYDNEY • TOKYO
Academic Press is an imprint of Elsevier

Elsevier Academic Press
525 B Street, Suite 1900, San Diego, California 92101-4495, USA
84 Theobald's Road, London WC1X 8RR, UK

This book is printed on acid-free paper. ∞

For all information on all Academic Press publications
visit our Web site at www.academicpress.com

ISBN: 0-12-182783-6

PRINTED IN THE UNITED STATES OF AMERICA
04 05 06 07 08 9 8 7 6 5 4 3 2 1

Table of Contents

Section I. Cooperative Binding

Contributors to Volume 379

Article numbers are in parentheses and following the names of contributors. Affiliations listed are current.

GARY K. ACKERS (3), *Department of Biochemistry and Molecular Biophysics, Washington University School of Medicine, St. Louis, Missouri 63110*

DOUG BARRICK (28), *Department of Biophysics, The Johns Hopkins University, Baltimore, Maryland 21218*

DOROTHY BECKETT (209), *Department of Chemistry and Biochemistry, University of Maryland, College Park, Maryland 20742*

ALBERTO BOFFI (55), *CNR Institute of Molecular Biology and Pathology, Department of Biochemical Sciences, University of Rome "La Sapienza," 0185 Rome, Italy*

FRED BREWER (107), *Department of Molecular Pharmacology, Albert Einstein College of Medicine, Bronx, New York 10461*

THOMAS BRITTAIN (64), *School of Biological Sciences, University of Aukland, Auckland, New Zealand*

E. SETHE BURGIE (3), *Department of Biochemistry and Molecular Biophysics, Washington University School of Medicine, St. Louis, Missouri 63110*

YI-DER CHEN (145), *Laboratory of Biological Modeling, National Institutes of Diabetes, Digestive and Kidney Diseases, National Intitutes of Health, Bethesda, Maryland 20892–5621*

EMILIA CHIANCONE (55), *CNR Institute of Molecular Biology and Pathology, Department of Biochemical Sciences, University of Rome "La Sapienza," 0185 Rome, Italy*

TARUN K. DAM (107), *Department of Molecular Pharmacology, Albert Einstein College of Medicine, Bronx, New York 10461*

HEINZ DECKER (81), *Institute for Molecular Biophysics, Johannes Gutenberg University, D-55128 Mainz, Germany*

HERMANN HARTMANN (81), *Institute for Molecular Biophysics, Johannes Gutenberg University, D-55128 Mainz, Germany*

CHIEN HO (28), *Department of Biological Sciences, Carnegie Mellon University, Pittsburgh, Pennsylvania 15213–2683*

JO M. HOLT (3), *Department of Biochemistry and Molecular Biophysics, Washington University School of Medicine, St. Louis, Missouri 63110*

JONATHAN A. LUKIN (28), *Affinium Pharmaceuticals, University Avenue, North Tower, Toronto, Ontario, Canada M5J1V6*

KATHLEEN S. MATTHEWS (188), *Department of Biochemistry and Cell Biology, Rice University, Houston, Texas 77005–1892*

LAWRENCE J. PARKHURST (235), *Department of Chemistry, University of Nebraska, Lincoln, Nebraska, 68588–0304*

SIDDHARTHA ROY (175), *Department of Biophysics, Bose Institute, Calcutta 700 054 India*

IRINA M. RUSSU (152), *Department of Chemistry, Wesleyan University, Middletown, Connecticut 06459*

FREDERICK P. SCHWARZ (128), *Center for Advanced Research in Biotechnology, National Institute of Standards and Technology, Rockville, Maryland 20850*

VIRGIL SIMPLACEANU (28), *Department of Biological Sciences, Carnegie Mellon University, Pittsburgh, Pennsylvania 15213–2683*

LISKIN SWINT-KRUSE (188), *Department of Biochemistry and Cell Biology, Rice University, Houston, Texas 77005–1892*

CONNIE S. YARIAN (3), *Department of Biochemistry and Molecular Biophysics, Washington University School of Medicine, St. Louis, Missouri 63110*

Preface

One of the most intriguing problems in biological energetics is that of cooperativity. From the discovery of cooperativity and allostery in hemoglobin 100 years ago (Bohr *et al.,* 1904)[1] to the characterization of cooperativity in a myriad of processes in modern times (i.e., transport, catalysis, signaling, assembly, folding), the molecular mechanisms by which energy is transferred from one part of a macromolecule to another continues to challenge us. Of course, the problem has many layers, as a molecule as "simple" and familiar as hemoglobin can simultaneously sense the chemical potential of each physiological ligand and adjust its interactions with the others accordingly. Ironically, the very allosteric intermediates that hold the structural and energetic secrets of cooperativity are the same whose populations are suppressed and, in many instances, largely obscured by the nature of cooperativity itself. Thus, innovative methodologies and techniques have been developed to address cooperative systems, many of which are presented in this volume *Energetics of Biological Macromolecules Part D* and its companion volume, *Part E*. The reader will observe remarkable similarities among the wide range of experimental strategies employed, attesting to fundamental issues inherent in all cooperative systems.

Jo M. Holt
Michael L. Johnson
Gary K. Ackers

---

[1]Bohr, C., Hasselbach, K. A., and Krogh, A. (1904) Ueber einen in biologischer Beziehung wichtigen Einfluss, den die Kohlensaurespannung des Blutes aufdessen Sauerstoffbindung ubt. *Skand. Arch. Physiol.* **16,** 402–412.

METHODS IN ENZYMOLOGY

VOLUME XXXV. Lipids (Part B)
Edited by JOHN M. LOWENSTEIN

VOLUME XXXVI. Hormone Action (Part A: Steroid Hormones)
Edited by BERT W. O'MALLEY AND JOEL G. HARDMAN

VOLUME XXXVII. Hormone Action (Part B: Peptide Hormones)
Edited by BERT W. O'MALLEY AND JOEL G. HARDMAN

VOLUME XXXVIII. Hormone Action (Part C: Cyclic Nucleotides)
Edited by JOEL G. HARDMAN AND BERT W. O'MALLEY

VOLUME XXXIX. Hormone Action (Part D: Isolated Cells, Tissues, and Organ Systems)
Edited by JOEL G. HARDMAN AND BERT W. O'MALLEY

VOLUME XL. Hormone Action (Part E: Nuclear Structure and Function)
Edited by BERT W. O'MALLEY AND JOEL G. HARDMAN

VOLUME XLI. Carbohydrate Metabolism (Part B)
Edited by W. A. WOOD

VOLUME XLII. Carbohydrate Metabolism (Part C)
Edited by W. A. WOOD

VOLUME XLIII. Antibiotics
Edited by JOHN H. HASH

VOLUME XLIV. Immobilized Enzymes
Edited by KLAUS MOSBACH

VOLUME XLV. Proteolytic Enzymes (Part B)
Edited by LASZLO LORAND

VOLUME XLVI. Affinity Labeling
Edited by WILLIAM B. JAKOBY AND MEIR WILCHEK

VOLUME XLVII. Enzyme Structure (Part E)
Edited by C. H. W. HIRS AND SERGE N. TIMASHEFF

VOLUME XLVIII. Enzyme Structure (Part F)
Edited by C. H. W. HIRS AND SERGE N. TIMASHEFF

VOLUME XLIX. Enzyme Structure (Part G)
Edited by C. H. W. HIRS AND SERGE N. TIMASHEFF

VOLUME L. Complex Carbohydrates (Part C)
Edited by VICTOR GINSBURG

VOLUME LI. Purine and Pyrimidine Nucleotide Metabolism
Edited by PATRICIA A. HOFFEE AND MARY ELLEN JONES

VOLUME LII. Biomembranes (Part C: Biological Oxidations)
Edited by SIDNEY FLEISCHER AND LESTER PACKER

VOLUME LIII. Biomembranes (Part D: Biological Oxidations)
Edited by SIDNEY FLEISCHER AND LESTER PACKER

VOLUME 72. Lipids (Part D)
Edited by JOHN M. LOWENSTEIN

VOLUME 73. Immunochemical Techniques (Part B)
Edited by JOHN J. LANGONE AND HELEN VAN VUNAKIS

VOLUME 74. Immunochemical Techniques (Part C)
Edited by JOHN J. LANGONE AND HELEN VAN VUNAKIS

VOLUME 75. Cumulative Subject Index Volumes XXXI, XXXII, XXXIV–LX
Edited by EDWARD A. DENNIS AND MARTHA G. DENNIS

VOLUME 76. Hemoglobins
Edited by ERALDO ANTONINI, LUIGI ROSSI-BERNARDI, AND EMILIA CHIANCONE

VOLUME 77. Detoxication and Drug Metabolism
Edited by WILLIAM B. JAKOBY

VOLUME 78. Interferons (Part A)
Edited by SIDNEY PESTKA

VOLUME 79. Interferons (Part B)
Edited by SIDNEY PESTKA

VOLUME 80. Proteolytic Enzymes (Part C)
Edited by LASZLO LORAND

VOLUME 81. Biomembranes (Part H: Visual Pigments and Purple Membranes, I)
Edited by LESTER PACKER

VOLUME 82. Structural and Contractile Proteins (Part A: Extracellular Matrix)
Edited by LEON W. CUNNINGHAM AND DIXIE W. FREDERIKSEN

VOLUME 83. Complex Carbohydrates (Part D)
Edited by VICTOR GINSBURG

VOLUME 84. Immunochemical Techniques (Part D: Selected Immunoassays)
Edited by JOHN J. LANGONE AND HELEN VAN VUNAKIS

VOLUME 85. Structural and Contractile Proteins (Part B: The Contractile Apparatus and the Cytoskeleton)
Edited by DIXIE W. FREDERIKSEN AND LEON W. CUNNINGHAM

VOLUME 86. Prostaglandins and Arachidonate Metabolites
Edited by WILLIAM E. M. LANDS AND WILLIAM L. SMITH

VOLUME 87. Enzyme Kinetics and Mechanism (Part C: Intermediates, Stereo-chemistry, and Rate Studies)
Edited by DANIEL L. PURICH

VOLUME 88. Biomembranes (Part I: Visual Pigments and Purple Membranes, II)
Edited by LESTER PACKER

VOLUME 89. Carbohydrate Metabolism (Part D)
Edited by WILLIS A. WOOD

VOLUME 90. Carbohydrate Metabolism (Part E)
Edited by WILLIS A. WOOD

VOLUME 91. Enzyme Structure (Part I)
Edited by C. H. W. HIRS AND SERGE N. TIMASHEFF

VOLUME 92. Immunochemical Techniques (Part E: Monoclonal Antibodies and General Immunoassay Methods)
Edited by JOHN J. LANGONE AND HELEN VAN VUNAKIS

VOLUME 93. Immunochemical Techniques (Part F: Conventional Antibodies, Fc Receptors, and Cytotoxicity)
Edited by JOHN J. LANGONE AND HELEN VAN VUNAKIS

VOLUME 94. Polyamines
Edited by HERBERT TABOR AND CELIA WHITE TABOR

VOLUME 95. Cumulative Subject Index Volumes 61–74, 76–80
Edited by EDWARD A. DENNIS AND MARTHA G. DENNIS

VOLUME 96. Biomembranes [Part J: Membrane Biogenesis: Assembly and Targeting (General Methods; Eukaryotes)]
Edited by SIDNEY FLEISCHER AND BECCA FLEISCHER

VOLUME 97. Biomembranes [Part K: Membrane Biogenesis: Assembly and Targeting (Prokaryotes, Mitochondria, and Chloroplasts)]
Edited by SIDNEY FLEISCHER AND BECCA FLEISCHER

VOLUME 98. Biomembranes (Part L: Membrane Biogenesis: Processing and Recycling)
Edited by SIDNEY FLEISCHER AND BECCA FLEISCHER

VOLUME 99. Hormone Action (Part F: Protein Kinases)
Edited by JACKIE D. CORBIN AND JOEL G. HARDMAN

VOLUME 100. Recombinant DNA (Part B)
Edited by RAY WU, LAWRENCE GROSSMAN, AND KIVIE MOLDAVE

VOLUME 101. Recombinant DNA (Part C)
Edited by RAY WU, LAWRENCE GROSSMAN, AND KIVIE MOLDAVE

VOLUME 102. Hormone Action (Part G: Calmodulin and Calcium-Binding Proteins)
Edited by ANTHONY R. MEANS AND BERT W. O'MALLEY

VOLUME 103. Hormone Action (Part H: Neuroendocrine Peptides)
Edited by P. MICHAEL CONN

VOLUME 104. Enzyme Purification and Related Techniques (Part C)
Edited by WILLIAM B. JAKOBY

VOLUME 105. Oxygen Radicals in Biological Systems
Edited by LESTER PACKER

VOLUME 106. Posttranslational Modifications (Part A)
Edited by FINN WOLD AND KIVIE MOLDAVE

VOLUME 125. Biomembranes (Part M: Transport in Bacteria, Mitochondria, and Chloroplasts: General Approaches and Transport Systems)
Edited by SIDNEY FLEISCHER AND BECCA FLEISCHER

VOLUME 126. Biomembranes (Part N: Transport in Bacteria, Mitochondria, and Chloroplasts: Protonmotive Force)
Edited by SIDNEY FLEISCHER AND BECCA FLEISCHER

VOLUME 127. Biomembranes (Part O: Protons and Water: Structure and Translocation)
Edited by LESTER PACKER

VOLUME 128. Plasma Lipoproteins (Part A: Preparation, Structure, and Molecular Biology)
Edited by JERE P. SEGREST AND JOHN J. ALBERS

VOLUME 129. Plasma Lipoproteins (Part B: Characterization, Cell Biology, and Metabolism)
Edited by JOHN J. ALBERS AND JERE P. SEGREST

VOLUME 130. Enzyme Structure (Part K)
Edited by C. H. W. HIRS AND SERGE N. TIMASHEFF

VOLUME 131. Enzyme Structure (Part L)
Edited by C. H. W. HIRS AND SERGE N. TIMASHEFF

VOLUME 132. Immunochemical Techniques (Part J: Phagocytosis and Cell-Mediated Cytotoxicity)
Edited by GIOVANNI DI SABATO AND JOHANNES EVERSE

VOLUME 133. Bioluminescence and Chemiluminescence (Part B)
Edited by MARLENE DELUCA AND WILLIAM D. MCELROY

VOLUME 134. Structural and Contractile Proteins (Part C: The Contractile Apparatus and the Cytoskeleton)
Edited by RICHARD B. VALLEE

VOLUME 135. Immobilized Enzymes and Cells (Part B)
Edited by KLAUS MOSBACH

VOLUME 136. Immobilized Enzymes and Cells (Part C)
Edited by KLAUS MOSBACH

VOLUME 137. Immobilized Enzymes and Cells (Part D)
Edited by KLAUS MOSBACH

VOLUME 138. Complex Carbohydrates (Part E)
Edited by VICTOR GINSBURG

VOLUME 139. Cellular Regulators (Part A: Calcium- and Calmodulin-Binding Proteins)
Edited by ANTHONY R. MEANS AND P. MICHAEL CONN

VOLUME 140. Cumulative Subject Index Volumes 102–119, 121–134

VOLUME 371. RNA Polymerases and Associated Factors (Part D)
Edited by SANKAR L. ADHYA AND SUSAN GARGES

VOLUME 372. Liposomes (Part B)
Edited by NEGAT DÜZGÜNEŞ

VOLUME 373. Liposomes (Part C)
Edited by NEGAT DÜZGÜNEŞ

VOLUME 374. Macromolecular Crystallography (Part D)
Edited by CHARLES W. CARTER, JR., AND ROBERT W. SWEET

VOLUME 375. Chromatin and Chromatin Remodeling Enzymes (Part A)
Edited by C. DAVID ALLIS AND CARL WU

VOLUME 376. Chromatin and Chromatin Remodeling Enzymes (Part B)
Edited by C. DAVID ALLIS AND CARL WU

VOLUME 377. Chromatin and Chromatin Remodeling Enzymes (Part C)
Edited by C. DAVID ALLIS AND CARL WU

VOLUME 378. Quinones and Quinone Enzymes (Part A)
Edited by HELMUT SIES AND LESTER PACKER

VOLUME 379. Energetics of Biological Macromolecules (Part D)
Edited by JO M. HOLT, MICHAEL L. JOHNSON, AND GARY K. ACKERS

VOLUME 380. Energetics of Biological Macromolecules (Part E)
Edited by JO M. HOLT, MICHAEL L. JOHNSON, AND GARY K. ACKERS

VOLUME 381. Oxygen Sensing (in preparation)
Edited by CHANDAN K. SEN AND GREGG L. SEMENZA

VOLUME 382. Quinones and Quinone Enzymes (Part B) (in preparation)
Edited by HELMUT SIES AND LESTER PACKER

VOLUME 383. Numerical Computer Methods, (Part D) (in preparation)
Edited by LUDWIG BRAND AND MICHAEL L. JOHNSON

VOLUME 384. Numerical Computer Methods, (Part E) (in preparation)
Edited by LUDWIG BRAND AND MICHAEL L. JOHNSON

VOLUME 385. Imaging in Biological Research, (Part A) (in preparation)
Edited by P. MICHAEL CONN

VOLUME 386. Imaging in Biological Research, (Part B) (in preparation)
Edited by P. MICHAEL CONN

Section I

Cooperative Binding

[1] Analyzing Intermediate State Cooperativity in Hemoglobin

By Gary K. Ackers, Jo M. Holt,

E. Sethe Burgie, and Connie S. Yarian

Introduction

The phenomenon of cooperative ligand binding is dramatically demonstrated in the distinctive sigmoidal O_2 binding curve of human hemoglobin (Hb), which facilitates efficient delivery of O_2 from the lungs to tissues undergoing oxidative metabolism. By the mid-1970s it had come to be believed by many that Hb had been conquered and its mechanism for cooperativity understood at the molecular level. Since then it has often been represented as a paradigm system for studying cooperativity and allostery. Although a great deal of physical and chemical characterization has been amassed in its study, the mechanism of cooperativity in Hb is currently undergoing major revision, based on the results of more modern experimental strategies implemented in the 1990s. The renovation is occurring at the most fundamental levels of understanding of how this molecule carries out its functions, including the number of quaternary structures the tetramer may assume and which of them dominates in solution, as well as which heme sites communicate with each other in the various stoichiometric and site-combinatorial forms of ligation. This chapter outlines the methodologies employed in the site-specific combinatorial investigation of the partially ligated Hb intermediates.

The methods for analyzing physical properties of distinct partially ligated intermediates have been used to develop a high-resolution understanding of cooperative ligand binding not available from the methods that yield average parameters and a low-resolution glimpse of cooperativity. The cornerstone of this approach has been the characterization of each of the ligation intermediates of Hb by linkage thermodynamics, independent of the structural and energetic models of cooperativity that dominated Hb research in previous decades. This chapter outlines the model-independent experimental approaches used to evaluate the contribution of each partially ligated Hb intermediate to the well-known sigmoidal O_2 binding curve.

Elements of the Tetrameric Hemoglobin Binding Isotherm

The Hb tetramer is composed of two α subunits and two β subunits structurally organized as two $\alpha\beta$ dimers. The dimer–dimer interface is polar and water filled, in contrast to the hydrophobic intradimer interface. Each subunit contains a heme moiety, the central Fe of which binds the physiological ligand O_2. The O_2 binding isotherm (Fig. 1) is typically quantitated in terms of the fractional saturation of all four binding sites, or $\overline{Y}$:

$$\overline{Y} = \frac{[\text{occupied binding sites}]}{[\text{total binding sites}]} = \frac{K_1x + 2K_2x^2 + 3K_3x^3 + 4K_4x^4}{4(1 + K_1x + K_2x^2 + K_3x^3 + K_4x^4)} \quad (1)$$

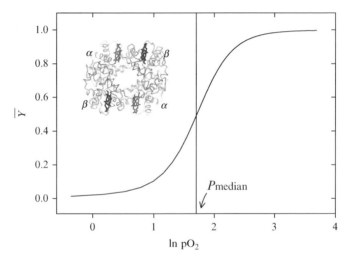

FIG. 1. Tetrameric Hb is composed of two α subunits and two β subunits, each containing a heme binding site. The binding curve of tetrameric Hb is illustrated here, calculated from experimentally determined Adair constants under standard *in vitro* conditions of pH 7.4, 21.5°, 0.1 M Tris, 0.1 M NaCl, 1 mM Na$_2$EDTA (total chloride concentration, 0.18 M).[1,2] The median ligand concentration, or P_{median}, is a function of the overall binding constant K_4:

$$(P_{\text{median}})^4 = \frac{1}{K_4}$$

It should be noted that this expression applies only to a nonassociating system, in this case a hypothetical solution composed only of Hb tetramers. For the actual solution of dimers and tetramers in equilibrium, a more complex relation applies. But this simple formulation serves to illustrate the nature of this fundamental thermodynamic parameter. The use of P_{median} is frequently substituted in the Hb literature by P_{50}, the partial pressure (or concentration) of ligand at $\overline{Y} = 0.5$, which is a convenient parameter, but has no direct relationship to binding energies. The P_{median} is the [O_2] at which the integrated area below the curve equals the integrated area above the curve, 5.3 mmHg under these conditions.

where K_i $(i = 1, \ldots, 4)$ is the equilibrium O_2 binding constant (K_1, K_2, K_3, and K_4 are collectively known as the Adair constants) and x is the partial pressure of O_2 or, more generally, the molar concentration of ligand X. The Adair constants are "product constants," that is, K_4 is the equilibrium constant for binding all four ligands (as opposed to the stepwise constant for binding only the fourth ligand).

Experimental determination of the four Adair constants directly from the O_2 binding curve yields well-resolved values for the first binding step, K_1, and the overall binding constant K_4. However, the second and third Adair constants, K_2 and K_3, are highly correlated.[3] Thus, many combinations of the two constants will give equally good fits to the O_2 binding curve. Even in the case of additional constraints, that is, in which a series of isotherms is measured over a range of different Hb concentrations, and then globally fit for the Adair constants, the errors associated with K_2 and K_3 are large.[1,2]

The Adair binding constants, K_i, are actually macroscopic, or composite, constants that do not pertain to any single binding reaction. Each Adair binding level (denoted by $i = 1, 2, 3,$ or 4) contains multiple tetrameric species that differ in the configuration of bound ligands (Fig. 2). A more detailed expression for $\overline{Y}$, in terms of microscopic constants, takes the form

$$\overline{Y} = \frac{(2k_{11} + 2k_{12})x + 2(2k_{21} + 2k_{22} + k_{23} + k_{24})x^2 + 3(2k_{31} + 2k_{32})x^3 + 4k_{41}x^4}{4[1 + (2k_{11} + 2k_{12})x + (2k_{21} + 2k_{22} + k_{23} + k_{24})x^2 + (2k_{31} + 2k_{32})x^3 + k_{41}x^4]}$$

(2)

where k_{ij} is the product microscopic binding constant for i ligands in configuration j.[4] To characterize each partially ligated intermediate of Hb, the microscopic constants k_{ij} must be measured. Values for k_{ij} cannot be determined from the Adair constants, as K_i is essentially a lower resolution constant than k_{ij} (Table I). In fact, there is no solution for Eq. (2) from the O_2 binding curve that uniquely identifies all nine k_{ij} ligand binding constants.

Assembly: Dimer–Tetramer Equilibrium

The previous sections have shown how the O_2 binding isotherm is composed of nine microscopic binding constants, k_{ij}. Even at this stage it is clear that the experimental approach to k_{ij} must extend beyond the O_2 binding isotherm. However, the complete description of the isotherm is still

[1] F. C. Mills, M. L. Johnson, and G. K. Ackers, *Biochemistry* **15**, 5350 (1976).
[2] A. H. Chu, B. W. Turner, and G. K. Ackers, *Biochemistry* **23**, 604 (1984).
[3] M. L. Johnson, H. R. Halvorson, and G. K. Ackers, *Biochemistry* **15**, 5363 (1976).
[4] G. K. Ackers, *Adv. Protein Chem.* **51**, 185 (1998).

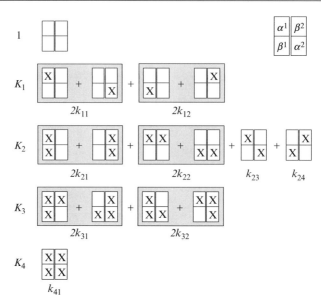

FIG. 2. The tetrameric Hb species configurations grouped according to macroscopic and microscopic ligand binding constants, K_i and k_{ij}, respectively. Placement of ligand is denoted by X. The ij designation describes the number (i) and configuration (j) of bound ligand. The paired tetramers shown in boxes are chemically indistinguishable isomers, and are shown here individually as each contributes to the statistical factor of two shown for the microscopic binding constant. The macroscopic constant is the sum of all contributing microscopic constants. For example, K_2, the constant for binding two ligands, $= 2k_{21} + 2k_{22} + k_{23} + k_{24}$.

more complex, because the contribution of dimer–tetramer assembly must be included. Although it is the Hb tetramer whose cooperative mechanism is of interest, the tetramer cannot be rigorously characterized in the absence of its component dimers. And the dimers exhibit distinctly different O_2 binding properties than do the tetramers, that is, the dimers bind O_2 with high affinity and no cooperativity.

The equilibrium between dimer and tetramer is typically expressed in terms of assembly:

$$2\alpha\beta \overset{{}^{ij}K_2}{\rightleftharpoons} \alpha_2\beta_2 \tag{3}$$

where ${}^{ij}K_2$ is the assembly equilibrium constant for the ij tetrameric species. The dimer population is enhanced as ligand concentration is increased, in fact, the assembly constant is remarkably sensitive to ij, varying over a 47,000-fold range from $10^{10}M^{-1}$ for deoxyHb to 10^5 for fully ligated Hb at pH 7.4 and $21.5°$.[1,3]

TABLE I

PARAMETERS OF HEMOGLOBIN OXYGEN BINDING CURVE

Lowest resolution	P_{median}	Median ligand concentration
	$\left.\begin{array}{l} K_1 \\ K_2 \\ K_3 \\ K_4 \end{array}\right\}$	Adair constants
Highest resolution	$\left.\begin{array}{l} k_{11} \\ k_{12} \\ k_{21} \\ k_{22} \\ k_{23} \\ k_{24} \\ k_{31} \\ k_{32} \\ k_{41} \end{array}\right\}$	Microstate constants

To derive an expression for the fractional saturation function $\overline{Y}$ that includes the contribution from dimers, it is useful to begin with the binding polynomial $\mathbf{Z}$, from which $\overline{Y}$ is derived. The binding polynomial is composed of (1) the sum of the concentrations of all forms of the tetramer TX_i relative to the unligated tetramer T, which serves as a reference species,

$$\mathbf{Z}_{\text{tetramer}} = \frac{[T] + [TX_1] + [TX_2] + [TX_3] + [TX_4]}{[T]}$$
$$= 1 + K_1 x + K_2 x^2 + K_3 x^3 + K_4 x^4 \tag{4a}$$

and (2) the concentrations of all forms of free dimer relative to unligated dimer D:

$$\mathbf{Z}_{\text{dimer}} = \frac{[D] + [DX_1] + [DX_2]}{[D]} = 1 + K_{D1} x + K_{D2} x^2 \tag{4b}$$

where K_{D1} and K_{D2} are the equilibrium constants for binding one and two ligands, respectively, to the free dimer.[5]

In general, the area under the binding curve ($\overline{Y}$ versus $\ln P_{O_2}$) gives the binding polynomial $\mathbf{Z}_{\text{Hb}}$, since

---

[5] G. K. Ackers and H. R. Halvorson, *Proc. Natl. Acad. Sci. USA* **71**, 4312 (1974).

$$\overline{Y} = \frac{1}{n}\frac{d\ln \mathbf{Z}_{Hb}}{d\ln x} \qquad (5)$$

where n is the number of binding sites per macromolecule (four in the case of tetrameric Hb, or two in the case of free dimer) and $\mathbf{Z}_{Hb} = \mathbf{Z}_{tetramer} + \mathbf{Z}_{dimer}$, which are related to each other by the assembly constant $^iK_2$. The concentration-dependent expression for $\overline{Y}$ in terms of $\mathbf{Z}_{tetramer}$, $\mathbf{Z}_{dimer}$, the total protein concentration P_t, and $^0K_2$ is given in Fig. 3.

Thus, a set of equilibrium binding parameters which completely describes the experimental O_2 isotherm of Hb is composed of tetrameric binding constants (nine microscopic k_{ij} values), dimeric binding constants K_{D1} and K_{D2}, and at least one dimer–tetramer assembly constant (typically that for deoxyHb, $^0K_2$). Given the large number of parameters at play, it is not surprising that many are highly correlated with one another and that there is no unique solution for all parameters.

$$\overline{Y} = \frac{\mathbf{Z}'_{dimer} + \mathbf{Z}'_{tetramer}\left[\left(-\mathbf{Z}_{dimer} + \sqrt{\mathbf{Z}^2_{dimer} + 4\,^0K_2\,\mathbf{Z}_{tetramer}\,[P_t]}\right)/4\mathbf{Z}_{tetramer}\right]}{\mathbf{Z}_{dimer} + \sqrt{\mathbf{Z}^2_{dimer} + 4\,^0K_2\,\mathbf{Z}_{tetramer}\,[P_t]}}$$

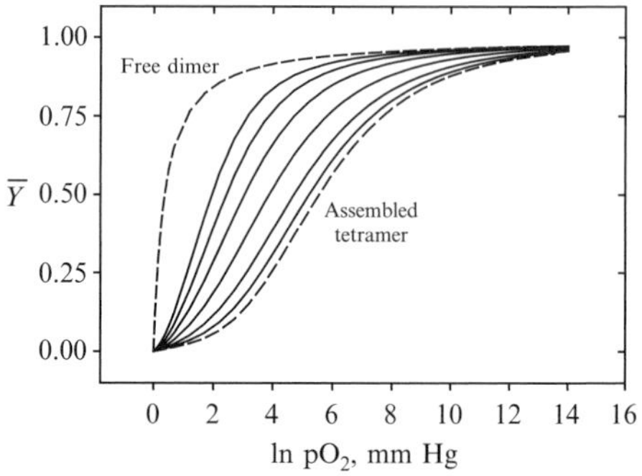

$$\mathbf{Z}_{dimer} = 1 + K_{D1}[O_2] + K_{D2}[O_2]^2$$

$$\mathbf{Z}'_{dimer} = K_{D1}[O_2] + 2K_{D2}[O_2]^2$$

$$\mathbf{Z}_{tetramer} = 1 + (2k_{11} + 2k_{12})[O_2] + (2k_{21} + 2k_{22} + k_{23} + k_{24})[O_2]^2 + (2k_{31} + 2k_{32})[O_2]^3 + (k_{41})[O_2]^4$$

$$\mathbf{Z}'_{tetramer} = (2k_{11} + 2k_{12})[O_2] + 2(2k_{21} + 2k_{22} + k_{23} + k_{24})[O_2]^2 + 3(2k_{31} + 2k_{32})[O_2]^3 + 4(k_{41})[O_2]^4$$

FIG. 3. The dependence of fractional saturation, $\overline{Y}$, with total protein concentration $[P_t]$. As $[P_t] \rightarrow \infty$, the $\mathbf{Z}_{tetramer}$ components of $\overline{Y}$ dominate, and as $[P_t] \rightarrow 0$, the $\mathbf{Z}_{dimer}$ components dominate. These two limiting cases are illustrated by the dashed isotherms. *In vitro*, the experimentally determined isotherm is a mixture of tetrameric and dimeric isotherms. The tetramer binding polynomial and its derivative are given in terms of microscopic binding constants to illustrate the contributions of each of the configurational isomers. In practice, concentration-dependent isotherm sets are analyzed with respect to the macroscopic Adair constants only.

Thermodynamic Linkage of Binding and Assembly

The interdependence between ligation and dimer to tetramer assembly, although expanding the Hb system to an additional level of complexity, also provides the means by which to approach the task of their deconvolution, in that assembly and ligation constants are linked to one another by the conservation property of Gibbs free energy, as originally recognized by Wyman.[6] The interdependence between ligand binding to tetramer, ligand binding to dimer, and assembly of dimers to tetramers is illustrated in Fig. 4. In an initial step toward formulating thermodynamic linkage for the partially ligated intermediates of Hb, Ackers and co-workers[5,7]

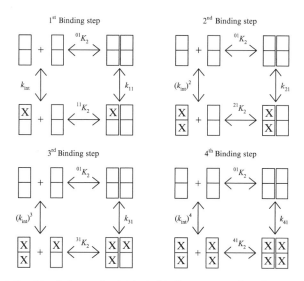

FIG. 4. Relationships between the intrinsic binding constant k_{int}, the tetrameric binding constant k_{ij}, and the assembly constant $^{ij}K_2$ for a representative subset of the ligand binding steps in Hb. The first ligand X binds to an α-subunit heme site in this example. The assembly constant $^{11}K_2$ is less than $^{01}K_2$, and k_{11} is less than k_{int}. In the second binding step, X is bound to a β-subunit heme site on the $\alpha\beta$ dimer that already carries a ligand on the α subunit. The thermodynamic box is drawn to show net binding from deoxy to doubly ligated tetramer to emphasize that the tetramer binding constant, k_{21}, is a product rather than a stepwise constant, even though binding of the two ligands is not simultaneous. In the third binding step, X is bound to a β subunit on the deoxy dimer, followed by the fourth step, resulting in completely ligated tetramer. The microscopic constant k_{41} is equivalent to the overall Adair binding constant K_4.

[6] J. Wyman, *Adv. Protein Chem.* **19**, 223 (1964).
[7] F. R. Smith and G. K. Ackers, *Proc. Natl. Acad. Sci. USA* **82**, 5347 (1985).

recognized that each ligand binding constant k_{ij} can be expressed in terms of a cooperativity constant $^{ij}k_c$. The cooperativity constant provides a measure of cooperative tetrameric binding ratioed against noncooperative intrinsic binding for each binding step:

$$^{ij}k_c = \frac{k_{ij}}{k_{int}^i} \tag{6}$$

where k_{int} is an intrinsic binding constant, set here as the value of the non-cooperative free dimer binding constant K_{D1} or K_{D2}. The nine microscopic tetramer binding constants can each be expressed in these terms:

$$
\begin{aligned}
k_{11} &= {}^{11}k_c(k_{int}) & k_{21} &= {}^{21}k_c(k_{int})^2 & k_{31} &= {}^{31}k_c(k_{int})^3 & k_{41} &= {}^{41}k_c(k_{int})^4 \\
k_{12} &= {}^{12}k_c(k_{int}) & k_{22} &= {}^{22}k_c(k_{int})^2 & k_{32} &= {}^{32}k_c(k_{int})^3 \\
& & k_{23} &= {}^{23}k_c(k_{int})^2 \\
& & k_{24} &= {}^{24}k_c(k_{int})^2
\end{aligned}
\tag{7}
$$

Thus, the binding polynomial $\mathbf{Z}_{Hb}$ effectively incorporates both dimeric and tetrameric binding constants into a single expression:

$$\mathbf{Z}_{Hb} = 1 + (2^{11}k_c + 2^{12}k_c)s + (2^{21}k_c + 2^{22}k_c + {}^{23}k_c + {}^{24}k_c)s^2 \\ + (2^{31}k_c + 2^{32}k_c)s^3 + {}^{41}k_cx^4 \tag{8}$$

where $s = (k_{int}x)$. The corresponding expression for $\overline{Y}$, as given in Eq. (2), in terms of cooperativity constants is then[4]

$$\overline{Y} = \frac{(2^{11}k_c + 2^{12}k_c)s + 2(2^{21}k_c + 2^{22}k_c + {}^{23}k_c + {}^{24}k_c)s^2 + 3(2^{31}k_c + 2^{32}k_c)s^3 + 4\,{}^{41}k_cx^4}{4[1 + (2^{11}k_c + 2^{12}k_c)s + (2^{21}k_c + 2^{22}k_c + {}^{23}k_c + {}^{24}k_c)s^2 + (2^{31}k_c + 2^{32}k_c)s^3 + {}^{41}k_cx^4]} \tag{9}$$

Of course, the four Adair cooperativity constants, ^iK_c, may be derived from the microscopic cooperativity constants to yield simpler expressions. However, the substitution of Adair constants for microscopic constants represents a loss of resolution in the system.

The cooperativity constant $^{ij}k_c$ can also be expressed as the cooperative free energy, $^{ij}\Delta G_c$, which is more typically used in high-resolution studies of Hb:

$$^{ij}\Delta G_c = -RT \ln {}^{ij}k_c \tag{10}$$

Since $^{ij}k_c$ is composed of both tetrameric and dimeric binding parameters, the cooperative free energy can also be expressed in those terms, that is, the tetrameric binding free energy minus the intrinsic binding free energy:

$$^{ij}\Delta G_c = -RT \ln \frac{k_{ij}}{k^i_{int}}$$
$$= -RT \ln k_{ij} + RT \ln k^i_{int}$$
$$^{ij}\Delta G_c = \Delta G_{ij} - \Delta G_{int}$$
(11)

Note: Cooperative free energy is a key conceptual element in this linkage thermodynamics approach. Consider the thermodynamic box for binding a single ligand, as shown in Fig. 5. The two-step pathway from deoxy dimers to ligated tetramer is an assembly step followed by ligation, or a ligation step before assembly. The total change in free energy is independent of path, so that

$$^{11}\Delta G_2 - {}^{01}\Delta G_2 = \Delta G_{11} - \Delta G_{int}$$
(12)

That is, the change in assembly free energy due to ligation must be equal to the change in binding free energy due to assembly. The change in assembly free energy on ligation therefore provides a measure of the cooperative free energy $^{ij}\Delta G_c$:

$$^{11}\Delta G_2 - {}^{01}\Delta G_2 = {}^{11}\Delta G_c$$
(13)

This thermodynamic linkage strategy permits $^{ij}\Delta G_c$ to be measured via assembly constants, providing experimental access to each of the microscopic

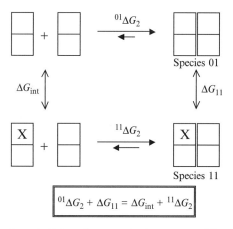

FIG. 5. The thermodynamic linkage between the free energy of ligation ΔG_{11} and dimer–tetramer assembly $^{11}\Delta G_2$, illustrated for binding the first ligand to an α subunit. The free energy changes for each reaction represent the net change in protein and ligand conformations on reaction, including bond formations and solvent rearrangements. The cooperative free energy for this binding step is expressed as $^{11}\Delta G_c = {}^{11}\Delta G_2 - {}^{01}\Delta G_2$ or equivalently as $^{11}\Delta G_c = \Delta G_{11} - \Delta G_{int}$.

binding steps not otherwise available from direct measurement of the O_2 binding isotherm.

Preparation of Parent Tetramers with Fixed Ligands

To measure the assembly constants for each partially ligated Hb intermediate, protocols have been developed for fixing the ligand in a subunit of interest in the symmetrically ligated tetrameric species, for example, tetramers in which both dimers have the same number and configuration of ligands. Because of this symmetry, species 01, 23, 24, and 41 do not disproportionate as a result of dimer exchange, and can be hybridized to form the asymmetric species. Most ligands of Hb, especially O_2, but also CO, are labile and move from heme site to heme site, creating a mix of partially ligated tetramers instead of a single (desired) species. Symmetrically ligated Hb tetramers are prepared with heme site analogs such as deoxy $Fe^{2+}/Fe^{3+}CN$, or Zn^{2+}/FeO_2 (in which the deoxy heme Fe^{2+} has been replaced with Zn^{2+}, which will not bind O_2), or $Co^{2+}FeCO^{4,8-10}$ (Fig. 6). The assembly free energies of the parent tetramers can be measured by employing techniques discussed below.

Parent Tetramer ΔG_2: Kinetic Approaches

Measurement of the dimer-to-tetramer assembly free energy of deoxy Hb is typically carried out by kinetic rather than equilibrium methods, because of the small abundance of free dimer in deoxyHb solutions at equilibrium. The dimer-to-tetramer association rate constant k_{on} is measured independently of the tetramer dissociation rate constant k_{off}, and both are used to calculate the equilibrium constant $^0K_2$.

Measurement of k_{on}: Dimer-to-Tetramer Assembly

To monitor the rate of dimer-to-tetramer assembly (k_{on}), a dilute oxygenated Hb solution is first prepared. Because the oxy tetramer is much less tightly associated than the deoxy tetramer, the oxyHb solution will contain significantly more free dimer than a deoxyHb solution of the same heme concentration. This dilute Hb solution is rapidly mixed with a buffered solution of dithionite, a powerful reducing agent that rapidly removes dissolved O_2 from both solvent and Hb. Deoxygenation results in

[8] G. McLendon and J. Feitelson, *Methods Enzymol.* **232,** 86 (1994).

[9] N. R. Naito, H. Huang, A. W. Sturgess, J. M. Nocek, and B. M. Hoffman, *J. Am. Chem. Soc.* **120,** 11256 (1998).

[10] R. Russo, L. Benazzi, and M. Perrella, *J. Biol. Chem.* **276,** 13628 (2001).

Human blood

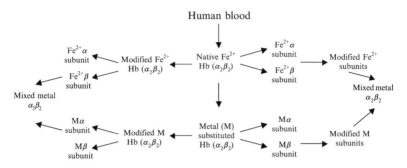

FIG. 6. Sequence for preparation of mixed metal Hb tetramer parents. Preparation begins with wild-type Hb. Substitution of native Fe^{2+} hemes (each subunit denoted by Fe in unshaded box) with hemes containing other metals, such as Mn^{2+} and Zn^{2+} (denoted by M) is performed on the tetramer. The left side of the flow chart shows the sequence for making a chemical or enzymatic modification to a specific residue on the tetramer followed by separation of α and β subunits. The right side of the flow chart shows the sequence for making a chemical or enzymatic modification to a specific residue on a purified α or β subunit, before recombining to form a mixed metal tetramer.

a transient mixture of deoxy free dimer and deoxy tetramer, apparent as a rapid increase in absorbance at 430 nm.[11] The free deoxy dimers then assemble to form tetramer, with an additional increase in A_{430}, apparently due to a difference in extinction coefficient between free dimers and tetramers. Measurement over a range of Hb concentrations[12-14] yields the second-order rate constant k_{on}, $1.1 \times 10^6 \ M^{-1} \ s^{-1}$ at $21.5°$ and pH 7.4. The value of k_{on} is independent of mutations, even those that significantly affect the dimer–tetramer equilibrium.[13] Thus, when the assembly equilibrium constant K_2 is altered, such as on ligation or mutation, it is the dissociation rate constant k_{off}, and not k_{on}, that has been affected.

Measurement of k_{off} *by Haptoglobin Dimer Trapping*

The rate constant for tetramer dissociation to free dimers, k_{off}, is determined by a method originally developed by Nagel and Gibson[15,16] using haptoglobin (Hp), a plasma protein that forms a complex with two free

[11] G. L. Kellett and H. Gutfreund, *Nature* **227,** 921 (1970).
[12] S. H. C. Ip, M. L. Johnson, and G. K. Ackers, *Biochemistry* **15,** 654 (1976).
[13] G. J. Turner, F. Galacteros, M. L. Doyle, B. Hedlund, D. W. Pettigrew, B. W. Turner, F. R. Smith, W. Moo-Penn, D. L. Rucknagel, and G. K. Ackers, *Proteins Struct. Funct. Genet.* **14,** 333 (1992).
[14] Y. Huang, M. L. Doyle, and G. K. Ackers, *Biophys. J.* **71,** 2094 (1996).
[15] R. L. Nagel and Q. H. Gibson, *J. Biol. Chem.* **246,** 69 (1971).
[16] R. L. Nagel and Q. H. Gibson, *Biochem. Biophys. Res. Commun.* **48,** 959 (1972).

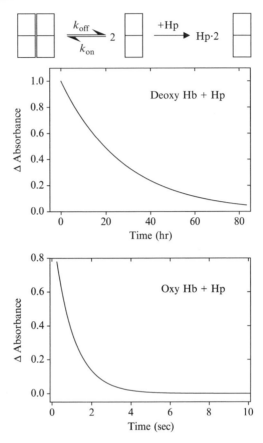

FIG. 7. Simulation of the absorbance change occurring on haptoglobin trapping of free dimer (shown in the schematic). A pseudo-first-order reaction with a rate similar to deoxygenated native Hb ($0.00001 \ s^{-1}$) and oxyHb ($1.0 \ s^{-1}$) illustrates the sensitivity of this reaction to the presence of ligands.

dimers. The Hp–dimer complex forms at a rate constant of approximately $5.5 \times 10^5 \ M^{-1} \ s^{-1}$ in an assembly that is essentially irreversible (the association equilibrium constant has been estimated[17] as at least $10^{15} \ M^{-1}$), thus effectively trapping the Hb dimer. With an excess of Hp relative to Hb, the overall rate-limiting step is Hb tetramer dissociation to free dimers (Fig. 7). The value of k_{off} is sensitive to protein modifications and ligand placement, varying from 1×10^{-5} to $1 \ s^{-1}$.[12,13] Given the consensus value for k_{on}, the

[17] P. K. Hwang and J. Greer, *J. Biol. Chem.* **255,** 3038 (1980).

Hp dimer-trapping technique can be used to determine dimer-to-tetramer assembly free energy, ΔG_2, ranging from -8.0 kcal/mol (corresponding to $k_{off} \approx 1 \text{ s}^{-1}$) to -14.4 kcal/mol (corresponding to $k_{off} \approx 2 \times 10^{-5} \text{ s}^{-1}$).[12]

This protocol is not applicable to all tetramers. Its use is limited to those Hb species for which the net reaction results in a significant change in absorbance. For the oxyHb tetramer (a stopped-flow experiment with a half-life of ~ 1 s), the absorbance change is too small to be of practical use, while carbon monoxyHb yields a large change in absorbance. DeoxyFe Hb reaction with Hp (best performed with manual mixing in a dual-beam spectrophotometer, half-life of ~ 7.5 h) occurs with a large absorbance change, while that with deoxyZn Hb has a much smaller, although still useful, absorbance change. Fluorescence can also potentially be used, but the complicating inner filter effect due to absorbance by the heme moiety greatly reduces the useable concentration range.

Parent Tetramer ΔG_2: Large Zone Chromatography

Hb tetramers with at least one ligand on each $\alpha\beta$ dimer are distinguished by their similar free energies of dimer-to-tetramer assembly, about -8 kcal/mol at pH 7.4, 21.5°.[18,19] At $\Delta G_2 = -8$ kcal/mol, the fraction of free dimer is large enough to permit its accurate determination by measuring the change in weight-average molecular weight with change in total Hb concentration (Fig. 8). This approach is not useful for the unligated Hb species, where $\Delta G_2 = -14.3$ kcal/mol, as the fraction of free dimer is too low for accurate measurement.

The weight-average molecular weight, or any similar parameter that reflects mass or size, can readily be obtained, from which values of $^{ij}\Delta G_2$ reproducible to within 0.1–0.2 kcal/mol can be derived. The most commonly employed technique is size-exclusion chromatography, but multiangle light-scattering techniques show great promise. Both protocols rely on the ability to measure the fraction of free dimer as total protein concentration changes.

Mole Fraction of Free Dimers

The fraction of free dimer is dependent on both the total protein concentration c_T and the dimer-to-tetramer association constant $^{ij}K_2$. To derive an expression for the fraction of dimer, the association constant is first defined:

[18] M. Perrella, L. Benazzi, M. A. Shea, and G. K. Ackers, *Biophys. Chem.* **35**, 97 (1990).
[19] G. K. Ackers, M. L. Doyle, D. Myers, and M. A. Daugherty, *Science* **255**, 54 (1992).

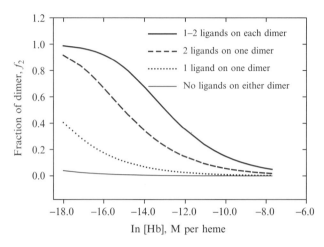

FIG. 8. The concentration dependence of the fraction of free dimers for different ligation configurations at pH 7.4, 21.5°. DeoxyHb exhibits the most negative assembly free energy (−14.3 kcal/mol) and the fraction of free dimer present, even at low Hb concentration, is always low. Tetramers with at least one ligand on each dimer have the weakest assembly free energy (−8.0 kcal/mol) and exhibit significant free dimer, particularly at lower concentrations. Most *in vitro* experiments on Hb are carried out in the 50–100 μM range (on a heme basis), which translates to −9 to −10 on a natural log scale.

$$^{ij}K_2 = \frac{[\text{tetramer}]}{[\text{dimer}]^2} = \frac{c_4}{(c_2)^2} \tag{14}$$

The concentration of tetramer, c_4, has terms of moles of tetramer per liter. Likewise, c_2 is expressed in moles of dimer per liter. The total concentration in moles of heme per liter is $c_T = 2c_2 + 4c_4$. The term $4c_4$ is therefore the concentration of hemes contributed by tetramer, and $2c_2$ is the concentration of hemes contributed by free dimer. The tetramer term can be substituted by the dimer term and c_T:

$$^{ij}K_2 = \frac{(c_T - 2c_2)}{(2c_2)^2}$$

Rearranging terms in order to solve the quadratic:

$$^{ij}K_2(2c_2)^2 = (c_T - 2c_2)$$
$$4^{ij}K_2c_2^2 + 2c_2 - c_T = 0$$

Yields, on substitution:

$$c_2 = \frac{-1 + \sqrt{1 + 4^{ij}K_2 c_T}}{4K_2}$$

Converting from concentration, on a molar heme basis, to mass fraction (the fraction of dimer hemes relative to all hemes), the mass fraction of dimer is

$$f_2 = \frac{2c_2}{c_T}, \quad \therefore \quad c_2 = \frac{f_2 c_T}{2}$$

$$f_2 = \frac{-1 + \sqrt{1 + 4^{ij}K_2 c_T}}{2^{ij}K_2 c_T} \tag{15}$$

Analytical Gel Permeation Chromatography

Gel chromatography techniques are based on the molecular size-dependent partitioning of solute molecules between solvent spaces within porous particles and the solvent space exterior to the particles. Ideally, this partitioning is dependent on molecular size and shape, with little or no influence from charge or surface properties. At equilibrium the distribution of macromolecular species is described by a partition isotherm that defines the relationship between the weight of protein Q_i inside the gel and the weight concentration of protein C (g/liter) in the void space exterior to the gel:

$$\sigma_j = \frac{Q_i}{V_i C} \tag{16}$$

where σ_j is the partition coefficient, defined as the amount of solute species distributed into the gel per unit internal volume V_i and external concentration C.[20] The dimensionless quantity σ_j is thus a measure of the extent of solute penetration for species j within the interior solvent regions of the gel.

For a sample of associating proteins undergoing reversible equilibration, such as the Hb tetramer and dimer, size-exclusion chromatography does not typically yield separate bands corresponding to free $\alpha\beta$ dimer and assembled $\alpha_2\beta_2$ tetramer. Because the association/dissociation reactions are in rapidly reversible exchange, a single band (or zone) is observed that migrates according to a concentration-dependent partition coefficient $\bar{\sigma}_w$, which is a weight average of the partition coefficients for the free dimer σ_2 and tetramer σ_4:

$$\bar{\sigma}_w = f_2\sigma_2 + f_4\sigma_4 \tag{17}$$

[20] G. K. Ackers, *Adv. Protein Chem.* **24**, 343 (1970).

where f_2 and f_4 are weight fractions of free dimers and tetramers, respectively. Since $f_4 = 1 - f_2$, $\bar{\sigma}_w$ can be expressed in terms of the dimer fraction f_2:

$$\bar{\sigma}_w = \sigma_4 + f_2(\sigma_2 - \sigma_4) \tag{18}$$

During chromatography, the ratio of free dimer and tetramer will adjust according to the law of mass action, so that $\bar{\sigma}_w$ is highly dependent on total protein concentration c_T. To establish a region of constant solute concentration c_T within the column, at which the desired association equilibrium constant will be determined, a sample solution is loaded in sufficient volume to maintain the same plateau concentration on eventual elution from the column. The leading and trailing boundaries of the large zone sample neither overlap nor interfere with one another, unlike the small zone experiment. Instead, each point within the plateau region yields a value of $\bar{\sigma}_w$ that is dependent on the plateau concentration. To determine $\bar{\sigma}_w$ experimentally from an elution profile, the centroid of the leading and/ or trailing edge of the plateau is determined, in a manner analogous to the measurement of P_{median}. Experimental procedure then consists of conducting a series of large zone experiments at different total protein concentration over a wide range of concentration. Measurement of $\bar{\sigma}_w$ versus plateau concentration (Fig. 9) may then be used to resolve K_2.[20–22]

Hybrid Formation from Parent Tetramers

Symmetric tetramers (Fig. 10) are used as parent species to generate the remaining partially ligated hybrid tetramers. Each hybrid tetramer carries ligands distributed asymmetrically between the two dimers of the Hb tetramer, distinguishing it from the symmetrically ligated parent tetramers. In general, hybridization is initiated simply by mixing two symmetrically ligated tetramer species, such as a parent AA and parent BB (Fig. 11). The free dimers A and B will recombine with like dimers to reform the parent tetramers, and will also combine with one another to form the hybrid tetramer AB. At equilibrium, three tetramers will be present in proportions dictated by the relative amounts of parent AA and parent BB initially mixed together and by the assembly equilibrium constants, $^{AA}K_2$, $^{BB}K_2$, and $^{AB}K_2$:

[21] E. Chiancone, L. M. Gilbert, G. A. Gilbert, and G. L. Kellett, *J. Biol. Chem.* **243**, 1212 (1968).

[22] R. Valdes, Jr., *Methods Enzymol.* **61**, 125 (1979).

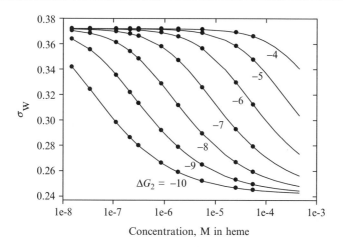

FIG. 9. Simulation of the dependence of the weight-average partition coefficient on protein concentration, illustrating the sensitivity of large zone chromatography for fully ligated tetramer under standard conditions. In practice, values for ΔG_2 have been measured for all three ligated parent tetramers under a wide range of conditions to ± 0.2 kcal/mol.[13,23–26] Typical concentration range for this experiment is four to six orders of magnitude.

$$^{AA}K_2 = \frac{[AA]}{[A]^2} \qquad ^{BB}K_2 = \frac{[BB]}{[B]^2} \qquad ^{AB}K_2 = \frac{2[AB]}{2[A]2[B]} = \frac{[AB]}{2[A][B]} \qquad (19)$$

The hybrid constant $^{AB}K_2$ is related to the parent constants by

$$^{AB}K_2 = \frac{[AB]\sqrt{^{AA}K_2\ ^{BB}K_2}}{2\sqrt{[AA][BB]}} \qquad (20)$$

Under conditions of high concentration in which the molar fraction of free dimer is insignificant, the total concentration of protein is the sum of the tetramer concentrations, $c_T = [AA] + [BB] + [AB]$, and the expression for $^{AB}K_2$ may be given[27] in terms of the molar fraction of tetramers:

$$\frac{^{AB}K_2}{\sqrt{^{AA}K_2\ ^{BB}K_2}} = \frac{f_{AB}}{2\sqrt{f_{AA}f_{BB}}} \qquad (21)$$

[23] G. K. Ackers, P. M. Dalessio, G. H. Lew, M. A. Daugherty, and J. M. Holt, *Proc. Natl. Acad. Sci. USA* **99,** 9777 (2002).

[24] M. A. Daugherty, M. A. Shea, and G. K. Ackers, *Biochemistry* **33,** 10345 (1994).

[25] Y. Huang and G. K. Ackers, *Biochemistry* **34,** 6316 (1995).

[26] Y. Huang, M. L. Koestner, and G. K. Ackers, *Biophys. Chem.* **64,** 157 (1997).

[27] V. J. LiCata, P. C. Speros, E. Rovida, and G. K. Ackers, *Biochemistry* **29,** 9771 (1990).

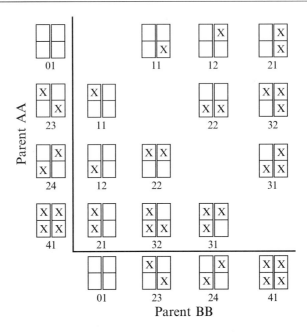

FIG. 10. Hybridization scheme for the preparation of Hb hybrid tetramers from the four Hb parent tetramers. Parent tetramers are distinguished from hybrid tetramers in that pure solutions of parents can be prepared and studied. Parents are prepared with heme site analogs that prevent ligands from binding to unligated or deoxy heme sites. For example, deoxyFe^{2+} hemes are replaced with Zn^{2+} hemes, the latter of which does not bind O$_2$ or CO. Hybrid tetramers cannot be studied alone in solution, as they are formed only in the presence of their two parent tetramers. Assembly free energy, ΔG_2, of deoxy parents is determined kinetically, ΔG_2 of ligated parents is determined by size-exclusion chromatography, and ΔG_2 of hybrids by subzero isoelectric focusing.

Values for $^{AA}K_2$ and $^{BB}K_2$ can be determined before hybridization, using methods described in the preceding sections, and the mole fractions of all three tetramers in the equilibrium hybridization mixture may be determined by isoelectric focusing (described below). The value for $^{AB}K_2$ is then determined from the fractional species abundances resulting from the hybridization experiment. The method employed for quantitation of the hybrid assembly constant then directly yields the assembly free energy, or $^{AB}\Delta G_2 = -RT \ln {}^{AB}K_2$.

Deviation Free Energy

Although it is reasonable to assume that the assembly free energy of the hybrid tetramer would be equal to the average of the assembly free energies of its parents,

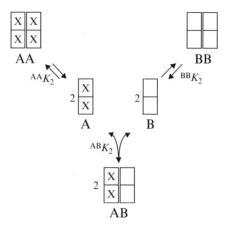

Fig. 11. Hybridization reaction scheme. Two parent tetramers are mixed, from which a hybrid tetramer is formed via the 2 dimer $\leftrightarrow\leftrightarrow$ tetramer assembly process. Parent tetramers are those that carry the same number of α- or β-subunit ligands on each dimer, that is, in which ligation is symmetrical. At equilibrium, the population of hybrid relative to that of the parents is dependent on the relative values of the three assembly constants.

$$^{AB}\Delta G_2 = \frac{^{AA}\Delta G_2 + {}^{BB}\Delta G_2}{2} \tag{22a}$$

the experimentally determined $^{AB}\Delta G_2$ typically is found to deviate from strict additivity, by an amount δ, called the deviation free energy:

$$\delta = {}^{AB}\Delta G_2 - \left(\frac{^{AA}\Delta G_2 + {}^{BB}\Delta G_2}{2}\right) \tag{22b}$$

The deviation free energy δ can be related to the molar fractions of tetramers by converting the ΔG terms to equilibrium constants and substituting Eq. (21)[27]:

$$\delta = -RT \ln \left[\frac{^{AB}K_2}{\sqrt{^{AA}K_2\,^{BB}K_2}}\right] = -RT \ln \left[\frac{f_{AB}}{2\sqrt{f_{AA}f_{BB}}}\right] \tag{23}$$

This expression is simplified when parent tetramers are mixed in equimolar amounts:

$$\delta = -RT \ln \left[\frac{f_{AB}}{1 - f_{AB}}\right] \tag{24}$$

Under these conditions, δ is dependent only on f_{AB}, and not on f_{AA} or f_{BB}. The experimental error for δ is therefore dependent only on the accuracy of f_{AB}, and the error in the independent determinations of $^{AA}K_2$ and $^{BB}K_2$ is not propagated into the free energy term δ.

Experimental Determination of δ

As shown in the previous section, the deviation free energy δ is dependent on the fractional concentrations f_{AA}, f_{BB}, and f_{AB}. These fractional concentrations of tetramers can be quantitatively evaluated by isoelectric focusing under conditions in which the dissociation of tetramer to free dimer is prevented. In quantitative cryogenic isoelectric focusing, qcIEF, a low-temperature quench (-25 to $-35°$) is applied to the hybridization mixture to prevent tetramer dissociation before loading onto the gel, which is then focused at the same low temperature.[27] The low temperature inhibits continuous perturbation of the dimer–tetramer equilibrium during isoelectric separation. Additional tetrameric stability may be provided by an anaerobic quench, in which O_2 is rapidly removed from heme sites by dithionite.[28] The deoxyHb tetramers thus generated have a half-life of dissociation to free dimer on the order of 7.5 h at $21.5°$, and even longer at subzero temperatures. The combination of an anaerobic quench at the temperature of hybridization (typically $21.5°$) followed by a cryogenic quench, and focusing performed anaerobically and cryogenically, provides assurance that the resolved Hb bands accurately represent the equilibrium species populations.

To increase resolution and separation distance between the two parent tetramers and the hybrid, one of the two parent tetramers carries an additional charge difference. The naturally occurring mutants HbC ($\beta6$ Glu $\rightarrow$ Lys) or HbS ($\beta6$ Glu $\rightarrow$ Val) are used for this purpose, since the surface amino acid modification in these mutants has no significant effect on their dimer–tetramer assembly free energies.[13,27]

Experimentally Accessible Range of δ

The preceding section describes the relationship of the deviation free energy, δ, to the experimentally observed equilibrium fraction of hybrid, f_{AB}. In turn, f_{AB} can be related to δ and the initial fractions of parents, f_{AA}^0 and f_{BB}^0. Solving Eq. (24) for f_{AB} yields

$$f_{AB} = 2e^{-\delta/RT} \sqrt{f_{AA} f_{BB}} \tag{25}$$

and

[28] G. K. Ackers, J. M. Holt, Y. Huang, Y. Grinkova, A. L. Klinger, and I. Denisov, *Proteins Struct. Funct. Genet.* **S4**, 23 (2000).

$$f_{AA} = f_{AA}^0 - \frac{f_{AB}}{2} \qquad f_{BB} = f_{BB}^0 - \frac{f_{AB}}{2} \qquad (26)$$

Substituting for f_{AA} and f_{BB}:

$$f_{AB} = 2e^{-\delta/RT} \sqrt{(f_{AA}^0 - f_{AB}/2)(f_{BB}^0 - f_{AB}/2)} \qquad (27)$$

Because $f_{AA}^0 + f_{BB}^0 = 1$, the expression simplifies:

$$f_{AB} = 2e^{-\delta/RT} \sqrt{f_{AA}^0 f_{BB}^0 - f_{AB}/2 + f_{AB}^2/4} \qquad (28)$$

Solving for f_{AB}:

$$f_{AB} = \frac{1 - \sqrt{1 - 4f_{AA}^0 f_{BB}^0 (1 - e^{2\delta/RT})}}{1 - e^{2\delta/RT}} \qquad (29)$$

This expression is equivalent to Eq. (9) in LiCata et al.[27] At $\delta = 0$, f_{AB} is simply

$$f_{AB} = 2f_{AA}^0 f_{BB}^0 \qquad (30)$$

The limit of detection of a gel band in the qcIEF procedure is typically 1% of total protein loaded. The limits of variation in the deviation free energy will therefore be set, at the lower limit, when $f_{AB} = 0.01$, and at the upper limit, when $f_{AA} = 0.01$ and $f_{BB} = 0.01$. The limits of detectability for δ are calculated, using Eq. (24), to be $+2.69$ to -2.28 kcal/mol under the conditions of these experiments. That is, a (hypothetical) 98-fold increase in f_{AB} would correspond to an increase of 5 kcal/mol in δ. This assumes that an equal amount of both parents were mixed in the hybridization (i.e., $f_{AA}^0 = f_{BB}^0 = 0.5$) (Fig. 12). As the ratio of f_{AA}^0/f_{BB}^0 varies from unity, the range of accessible δ values tightens dramatically. Once the initial ratio of parent species reaches 6, it becomes impossible to accurately quantitate hybrid species with $\delta = 0$, the most common condition.

Hybrid Tetramer ΔG_2: Kinetic Approaches

Although primarily used in determining ΔG_2 values for isolated symmetric parent tetramers (species 01, 23, 24, and 41; see Fig. 2), kinetic trapping of dimers with Hp can be applied to asymmetric tetramers as well, that is, tetramers whose component $\alpha\beta$ dimers are unequally ligated (species 11, 12, 21, 22, 31, and 32). However, asymmetric species must be measured along with both parent tetramers also present in the solution. The tetramer dissociation rate constants for each of the asymmetric hybrids were determined for the deoxyFe^{2+}/Fe^{3+}CN analog system, in which tetramer-to-dimer

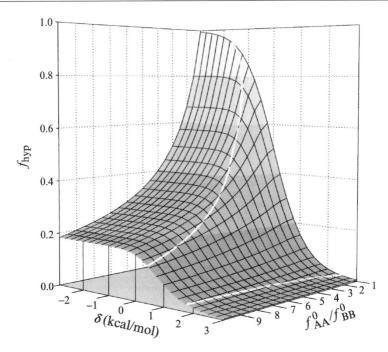

FIG. 12. Variation and limits on quantitation of the qcIEF experiment. The fraction of hybrid at equilibrium, f_{hyb}, is dependent on deviation free energy δ and molar ratio of parent species at time zero, f^0_{AA}/f^0_{BB}. The gel band density is linear over a 35-fold range of total Hb concentration, illustrated here with the solid white lines. Only the area between the white lines is experimentally accessible under these conditions.

dissociation rate constants ranged from 0.84×10^{-5} to $2.1 \times 10^{-5}\,\mathrm{s}^{-1}$ within a single experiment.[7] Data from stopped-flow and manual mixing methods were combined to determine all three tetramer dissociation rate constants:

$$\Delta Abs = Abs_\infty + b_1 e^{-k_{AA}t} + b_2 e^{-k_{AB}t} + b_2 e^{-k_{BB}t} \tag{31}$$

where the change in absorbance with time, ΔAbs, is a function of each k_{off} value (k_{AA} and k_{BB} for parents, k_{AB} for the hybrid).

Hybrid Tetramer ΔG_2: Analytical Gel Chromatography

Application of large zone chromatography to the determination of ΔG_2 values for those Hb species bearing at least one ligand on each dimer is typically limited to parent species that can be studied pure in solution, for example, species 41 (fully ligated), species 23 (ligands on both α subunits), and species 24 (ligands on both β subunits). However, the weight-average partition coefficient can also be determined for hybrid mixtures, such as for the singly ligated species.[14] A particularly elegant variation on this theme is the differential large zone technique, in which a pure solution of one Hb is layered over a pure solution of another Hb.[21,29,30] Large zone chromatography has also found applications in other associating systems,[31–34] and a faster version of the method that uses less protein has been proposed.[35]

Model-Independent Distribution of ΔG_c among the Hb Intermediates

The methodological strategy outlined in this chapter has produced binding data for each of the partially ligated intermediates of Hb, using multiple heme site analog systems. These binding data are in terms of $^{ij}\Delta G_c$, the cooperative free energy of ligation, in which the contribution of free dimer is explicitly deconvolved from that of assembled tetramer for each ij Hb intermediate. That is, $^{ij}\Delta G_c$ is the free energy that must be added to the intrinsic ΔG of ligation in order to give the ΔG of ligation to the tetramer. As is evident from a representative sampling of this data (summarized in Table II), $^{ij}\Delta G_c$ is typically positive, meaning that ligand binding to the tetramer is less favorable than to the free dimer. Because ΔG_c is usually positive, it is sometimes referred to as a "penalty."

Each heme site analog, whether an analog of a ligated hemesite (such as $Fe^{3+}CN$) or of a deoxy heme site (such as Co^{2+}), perturbs Hb conformation or subunit interactions to some extent. This perturbation is most commonly observed to decrease $^i\Delta G_c$ relative to known values for O_2 binding (i.e., the first binding step and the overall binding equilibrium). Although the degree and direction of perturbation due to analog substitution is variable, the pattern of ΔG_c distribution among the Hb intermediates is independent of the analog employed: tetramers with ligand(s) on both dimers within the tetramer have ΔG_c values that differ significantly from tetramers with ligand(s) on only one dimer. This dimer-based bias is particularly evident when comparing ΔG_c values for doubly ligated species (species 21, 22, 23, and 24). Regardless of the type of heme site analog, species 21, which

[29] G. A. Gilbert, *Nature* **212,** 296 (1966).
[30] M. Gattoni, M. C. Piro, A. Boffi, W. S. Brinigar, C. Fronticelli, and E. Chiancone, *Arch. Biochem. Biophys.* **386,** 172 (2001).
[31] S. Bandyopadhyay, C. Mukhopadhyay, and S. Roy, *Biochemistry* **35,** 5033 (1996).
[32] P. J. Darling, J. M. Holt, and G. K. Ackers, *Biochemistry* **39,** 11500 (2000).
[33] N. Funasaki, S. Ishikawa, and S. Neya, *Langmuir* **16,** 5584 (2000).
[34] O. Jaren, S. Harmon, A. F. Chen, and M. A. Shea, *Biochemistry* **39,** 6881 (2000).
[35] E. Nenortas and D. Beckett, *Anal. Biochem.* **222,** 366 (1994).

TABLE II
Use of Heme site Analogs in Studies of Hemoglobin Intermediates

			ΔG_c for heme site analogs (unligated/ligated)			
i	ij	Ligand (X) configuration	$Fe^{2+}/Fe^{3+}CN$	Fe^{2+}/Mn^{3+}	$Co^{2+}/Fe^{2+}CO$	Zn^{2+}/FeO_2
0	01	$\alpha\beta : \alpha\beta$	0	0	0	0
1	11	$\alpha X\beta : \alpha\beta$	3.1	2.9	1.5	2.9
	12	$\alpha\beta X : \alpha\beta$	3.3	3.7	2.0	2.9
2	21	$\alpha X\beta X : \alpha\beta$	4.0	3.4	2.1	5.1
	22	$\alpha X\beta : \alpha\beta X$	6.4	6.6	3.0	6.6
	23	$\alpha X\beta : \alpha X\beta$	6.1	6.8	3.1	6.8
	24	$\alpha\beta X : \alpha\beta X$	6.4	6.2	3.2	6.5
3	31	$\alpha X\beta X : \alpha\beta X$	6.3	6.5	3.0	6.9
	32	$\alpha X\beta X : \alpha X\beta$	6.2	6.5	3.1	6.9
4	41	$\alpha X\beta X : \alpha X\beta X$	6.0	6.9	2.5	6.3

has both ligands on one dimer within the tetramer, has a smaller ΔG_c value than do species 22, 23, and 24, which have one ligand on each dimer.

Application of the thermodynamic linkage approach also permits dissection of heterotropic effects, which to date has shown that those Hb tetramers with $Fe^{3+}CN$ ligand(s) on only one dimer (species 11, 12, and 21) have pH, temperature, and [NaCl] dependencies similar to those of the fully unligated tetramer.[4] The use of hybridization techniques permits the study of mutant hybrid tetramers, in which one dimer carries a mutation (or chemical modification), while the other dimer remains wild type.[23] A series of mutations using the $Fe^{3+}CN$ analog system has demonstrated the presence of autonomous function between the two dimers that make up the Hb tetramer. That is, the ΔG_c values for binding ligand to the modified dimer are similar to those measured for the classic, doubly modified Hb tetramer, whereas ΔG_c values for binding ligand to the wild-type dimer are similar to the wild-type Hb tetramer (Fig. 13).

Model of Dimer-Based Cooperativity in Human Hb

Although perturbations due to analog substitution at the heme site are always present, including ligand exchange artifacts in species 21 of the $Fe^{3+}CN$ system, the characteristic distribution of ΔG_c values among the 10 Hb species is remarkably self-consistent over a range of solution conditions and mutations.[19] The $\alpha\beta$ dimer is clearly capable of functioning as a

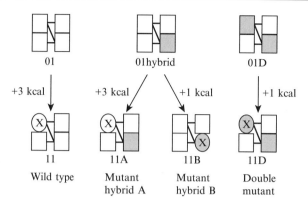

FIG. 13. Dimer autonomy within the Hb tetramer. Binding the first ligand (shown here as binding ligand X to an α-subunit heme site, designated species 11) occurs with a cooperative free energy of 3 kcal/mol in wild-type Hb. The same ligation process in a mutant Hb (the double mutant, with both α subunits carrying the modification, indicated by shading) exhibits a decreased ΔG_c of only 1 kcal/mol. In the mutant hybrid, ligation can occur either to the wild-type dimer (species 11A) or to the mutant dimer (species 11B). Ligation to the wild-type dimer occurs with a wild-type ΔG_c, whereas ligation to the mutant dimer occurs with a double mutant ΔG_c.

cooperative autonomous unit within the tetramer (Fig. 14). Ligand binding by one dimer cannot, however, be observed independently of its partner dimer in the O_2-binding isotherm, which reports only average binding constants for a mixture of intermediate ligation states. Thus, dimer functionality within the tetramer is directly observed only when the contributions of all intermediate ligation states are measured.

Summary

A complete thermodynamic description of O_2 binding to Hb includes ligand-binding constants for all tetrameric intermediates, ligand-binding constants for the free $\alpha\beta$ dimers, and assembly constants for dimer-to-tetramer assembly for all tetrameric intermediates. Even an array of globally fit O_2-binding isotherms cannot be deconvolved to yield unique values for these constants. This chapter has described an approach to this problem employing thermodynamic linkage analysis to hybrid Hbs whose ligand configurations have been fixed using nonlabile heme site analogs. This methodological strategy has revealed that the dimers within the Hb tetramer exhibit autonomous cooperativity of ligand binding. The functionality of the dimer within the tetramer has essentially been hidden in the classic experimental approaches to the study of cooperativity in Hb, and is

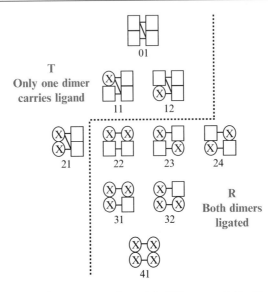

FIG. 14. Illustration of the symmetry rule model of Hb cooperativity. Ligand binding to the dimer within the tetramer occurs with positive cooperativity, even in the absence of significant quaternary change. However, when ligands are bound, in any combination, to both dimers within the tetramer, a quaternary switch occurs.

revealed only by a site-specific combinatorial evaluation of the partially ligated intermediates. As the era of macromolecular system analysis continues to develop, this kind of revelation may be expected to emerge with increasing frequency in many other macromolecular systems.

[2] Nuclear Magnetic Resonance Spectroscopy in the Study of Hemoglobin Cooperativity

By Doug Barrick, Jonathan A. Lukin, Virgil Simplaceanu, and Chien Ho

Introduction

Human normal adult hemoglobin (Hb A) is a classic system for understanding the structural basis of a complex set of cooperative, linked interactions and reactivities.[1,2] These interactions include homotropic effects such as the positively cooperative binding of molecular oxygen (O_2), and heterotropic effects such as the linkage between O_2 binding and the binding of effectors such as hydrogen ions (H^+), organic phosphates such as 2,3-bisphosphoglycerate (2,3-BPG) and inositol hexaphosphate (IHP), and chloride ions (Cl^-) to sites distributed throughout the tetramer. Although the thermodynamics of the ligand interactions described above have been quantified in considerable detail, the atomic structural details by which these ligands interact have not been firmly established. This chapter focuses on the use of nuclear magnetic resonance (NMR) to study the relationships between structure and ligand reactivity in Hb A.

Human Hemoglobin

Hb A exists as a tetramer containing two α subunits and two β subunits (Fig. 1). Each subunit contains a buried heme group to which a single O_2 molecule binds. This binding reaction involves direct axial coordination of each O_2 molecule to the ferrous iron of each of the heme groups. In addition, coordination can occur between the ferrous form of the heme iron and a number of other ligands (collectively referred to as distal ligands), including carbon monoxide (CO), nitric oxide (NO), and alkyl isocyanides (RNCs), and between the ferric form of the heme iron and water, cyanide (CN^-), azide (N_3^-), and fluoride ions. The four heme pyrrole nitrogens occupy four axial coordination sites to the iron, and an invariant "proximal" histidine from the "F" helix[*] of each subunit occupies the other axial coordination site, binding on the opposite face of the heme from the distal ligand.

Subunit assembly and quaternary structure have long been known to play important roles in tuning Hb affinity and in coupling the binding reactions between distant sites, as demonstrated by the dependence of the oxygen-binding curve on Hb A concentration. As Hb concentration is decreased (and as a result the stoichiometry shifts from $\alpha_2\beta_2$ tetramers to $\alpha\beta$ dimers), the oxygen-binding curve shifts to lower oxygen concentration,[3] indicating that tetramer formation diminishes reactivity. The response of quaternary structure within the $\alpha_2\beta_2$ tetramer to oxygen binding, an

[1] E. Antonini and M. Brunori, "Hemoglobin and Myoglobin in Their Reactions with Ligands." North-Holland Publishing, Amsterdam, 1971.

[2] J. Wyman and S. Gill, "Binding and Linkage: Functional Chemistry of Biological Macromolecules." University Science Books, Mill Valley, CA, 1990.

[3] F. C. Mills, M. L. Johnson, and G. K. Ackers, *Biochemistry* **15,** 5350 (1976).

[*] In the β subunits of hemoglobin (and in myoglobin), the helices are designated by lettering from A to H, making the F helix the sixth helix from the amino terminus. However, in the α subunits of hemoglobin, the D helix is missing, making the F helix the fifth helix from the amino terminus.[37]

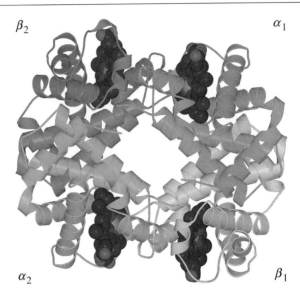

FIG. 1. The crystal structure of Hb A. α Subunits are shown in green and turquoise, and the β subunits are shown in yellow and orange. The $\alpha_1\beta_1$ dimer is on the right; the $\alpha_2\beta_2$ dimer is on the left. The four heme groups are shown as van der Waals spheres. Coordinates are from pdb file 1A3N. This image was prepared using MOLSCRIPT[4] and Raster3D.[5] (See color insert.)

important component of the allosteric mechanism in Hb A, was first suggested in experiments by Haurowitz, in which crystals of deoxy-Hb A shattered and then reformed in a different crystal lattice on exposure to oxygen.[6] The idea that Hb allostery may arise from the linkage of quaternary structure changes to the binding of distal ligands and second-site effectors was an important component of the concerted, two-state model developed by Monod, Wyman, and Changeux[7] and, to a lesser extent, of the stepwise model of Koshland, Nemethy, and Filmer.[8]

The first high-resolution model of how distal ligand binding is linked to quaternary structure changes came from Perutz's X-ray studies of the crystal structures of Hb A in the deoxy and metaquo forms.[9] A careful comparison of these structures led Perutz to propose a mechanism by which

[4] P. J. Kraulis, *J. Appl. Crystallogr.* **24,** 946 (1991).

[5] E. A. Merritt and D. J. Bacon, *Methods Enzymol.* **277,** 505 (1997).

[6] F. Haurowitz, *Z. Physiol. Chem.* **254,** 266 (1938).

[7] J. Monod, J. Wyman, and J. P. Changeux, *J. Mol. Biol.* **12,** 88 (1965).

[8] D. E. Koshland, Jr., G. Nemethy, and D. Filmer, *Biochemistry* **5,** 365 (1966).

[9] M. F. Perutz, *Nature* **228,** 726 (1970).

binding of distal ligands to heme sites could be coupled to one another.[9] In this mechanism, the binding of oxygen moves the heme iron to a more distal position, that is, into the plane of the heme, which repositions the proximal histidine, the F helix, and a region of the polypeptide that makes intersubunit contacts.[9–11] In addition, several salt bridges were observed to rupture in the transition from deoxy (T) to distally ligated (R) Hb A, and were originally proposed to be the sites of hydrogen ion binding that linked to oxygen binding by modulating the observed quaternary structure change.[9] A comparison of the quaternary structures of deoxy and distally ligated Hb crystal structures revealed significant rotation of one $\alpha\beta$ dimer with respect to the other ($\alpha_1\beta_1$ with respect to $\alpha_2\beta_2$), resulting in a large-scale rearrangement of the $\alpha_1\beta_2$ (and the symmetry-related $\alpha_2\beta_1$) interface.[10–12] In contrast, the $\alpha_1\beta_1$ (and symmetry-related $\alpha_2\beta_2$) interfaces were found to be structurally insensitive to distal ligand binding, consistent with the idea that the major quaternary transition occurs between the $\alpha_1\beta_1$ and $\alpha_2\beta_2$ dimers, but not within them.[12]

After more than 30 years of hemoglobin research, many aspects of Perutz's initial structural description of the allosteric transition have endured, although several aspects have faced challenges. The importance of the covalent linkage between the proximal histidines and the F helices has been experimentally confirmed by combining site-directed mutagenesis and chemical rescue to replace these histidines with free imidazole groups.[13,14] Advances in our understanding of hemoglobin allostery have been obtained from both structural and thermodynamic studies. X-ray crystallography has provided a detailed picture of the various structures of Hb A at increased resolution, in combination with different distal ligands, and in different crystal lattices. One important finding relating to the original Perutz analysis of Hb A crystal structures is that distally ligated forms of Hb A appear to adopt a variety of quaternary structures depending on crystallization conditions.[15–17] Whereas Perutz's "R" state structure was crystallized from high concentrations of ammonium sulfate, crystallization with polyethylene glycol under low-salt conditions produces a markedly different quaternary structure, referred to as "R2,"[16] which differs from

[10] A. Warshel, *Proc. Natl. Acad. Sci. USA* **74**, 1789 (1977).

[11] B. R. Gelin and M. Karplus, *Proc. Natl. Acad. Sci. USA* **74**, 801 (1977).

[12] J. Baldwin and C. Chothia, *J. Mol. Biol.* **129**, 175 (1979).

[13] D. Barrick, N. T. Ho, V. Simplaceanu, F. W. Dahlquist, and C. Ho, *Nat. Struct. Biol.* **4**, 78 (1997).

[14] D. Barrick, N. T. Ho, V. Simplaceanu, and C. Ho, *Biochemistry* **40**, 3780 (2001).

[15] T. C. Mueser, P. H. Rogers, and A. Arnone, *Biochemistry* **39**, 15353 (2000).

[16] M. M. Silva, P. H. Rogers, and A. Arnone, *J. Biol. Chem.* **267**, 17248 (1992).

[17] F. R. Smith and K. C. Simmons, *Proteins* **18**, 295 (1994).

the R structure both at the $\alpha_1\beta_2$ interface and in the overall orientation of the $\alpha_1\beta_1$ and $\alpha_2\beta_2$ dimers. Evaluation of various structural features of Hb A in the T, R, and R2 quaternary structures indicates that R is intermediate between T and R2.[18] Additional crystallographic studies of the fully ligated state of Hb A support the idea that the quaternary structure of the fully ligated state is plastic, being able to adopt a variety of quaternary interactions depending on pH, precipitant, and primary sequence.[15]

The structural heterogeneity of the different fully ligated Hb A crystal structures raises the question of "which R structure" is most appropriate to describe the quaternary structure of distally ligated Hb A in solution, if in fact the quaternary structure of Hb A can be represented using a single crystal structure. As pointed out by Arnone and co-workers, the observed heterogeneity indicates that different structures are energetically accessible and, thus, each structure should be substantially populated in solution.[15,16] Whether the observation of different structures under different crystallization conditions represents a population bias in solution that is imparted by different conditions, or whether different crystallization conditions simply stabilize one set of crystal lattice interactions over another, is not clear. To resolve these issues, a method is needed that can determine structure in solution under a variety of conditions. High-resolution multinuclear NMR spectroscopy appears to be uniquely suited to provide this structural information.

Scope of This Chapter

In this chapter, we focus on advances in the analysis of the structure–function relationships of Hb A, using NMR spectroscopy. Although Hb A is relatively large by NMR standards, NMR studies of Hb A have provided considerable insight into structure–function relationships. This insight has resulted from the structural features of Hb A that provide characteristic and unusually resolved chemical shifts in one-dimensional (1D) ^{1}H NMR, and also from the application of heteronuclear labeling strategies for production of protein samples suitable for high-resolution heteronuclear multidimensional spectroscopy. Several reviews have been published that describe methods, assignments, and structural information obtained from 1D ^{1}H NMR spectroscopy.[19,20] Therefore, we provide here only a brief overview of the methods and mechanisms that facilitate structure–function studies using 1D ^{1}H NMR and describe instead the high-resolution

[18] R. Srinivasan and G. D. Rose, *Proc. Natl. Acad. Sci. USA* **91**, 11113 (1994).

[19] C. Ho, *Adv. Protein Chem.* **43**, 153 (1992).

[20] J. A. Lukin and C. Ho, *in* "Methods in Molecular Medicine" (R. L. Nagel, ed.), Vol. 82, pp. 1–19. Humana Press, Totowa, NJ, 2003.

multidimensional multinuclear NMR spectroscopic methods applied to studies of isotopically labeled Hb A. This description is followed by an outline of expression methods and isotopic labeling strategies for the Hb A tetramer. In the next section, we present several contributions that NMR spectroscopy has made to understand the relation between structure and reactivity in this paradigm of allosteric interaction. We include results from both recent 1D ^{1}H NMR studies of recombinant hemoglobins (rHBs) bearing sequence substitutions and results from high-resolution multinuclear studies.

Basic Types of Information Available from NMR Spectroscopy of Hb A

In principle, NMR spectroscopy of proteins provides a powerful means to determine structure, conformational dynamics, and the short- and long-range response to distal ligand binding in Hb A. This is because, unlike other forms of solution spectroscopy, NMR directly detects each nucleus in the protein that has a spin of 1/2. Moreover, signals from individual nuclei are much sharper than the range of frequencies over which they are distributed, allowing for resolution of a large number of nuclear signals in a single spectrum. For proteins, the abundant ^{1}H nuclei have a spin of 1/2, and form the cornerstone of protein NMR spectroscopy.

In practice, several factors prevent simple 1D NMR spectroscopy from providing direct information about each (or even most) of the protons in a protein the size of Hb A. First, ^{1}H nuclei in proteins are numerous (on the order of several thousand), resulting in extreme spectral overlap of signals compared with small molecules and peptides. Second, because proteins as large as hemoglobin tumble relatively slowly in solution, resonances become broad, worsening the spectral overlap problem. Despite these problems, Hb possesses several structural features and interactions that partly offset these limiting factors, and have permitted critical information about ligand binding and allosteric changes to be obtained from simple 1D ^{1}H NMR spectroscopy.

One feature that has allowed structural and functional information to be obtained from 1D ^{1}H NMR is the presence of heme iron atoms in each of the four subunits. Depending on ligation and oxidation states of the heme irons, the d electrons of the irons can exist in a variety of different paramagnetic spin states (5/2 for high-spin ferric iron with a weak-field axial ligand such as water, 1/2 for low-spin ferric iron with a strong-field axial ligand such as CN^-, and 2 for high-spin ferrous iron without an axial ligand, as in deoxy-HbA). These paramagnetic complexes create strong local anisotropic magnetic fields and, as a result, produce large changes in chemical shift and relaxation properties of neighboring ^{1}H nuclei. These "hyperfine" interactions can provide highly detailed structural information about ^{1}H atoms of the heme and the residues in the heme pockets. The use of the hyperfine interaction to obtain structural information in Hb A has been reviewed by Ho.[19]

Changes in spin state of the iron d electrons that accompany ligand binding also serve as a useful subunit-specific (α versus β) probe of ligand binding because the large changes in the 1D ^{1}H NMR spectrum that result from these spin-state changes (e.g., from $S = 2$ for deoxy-Hb to $S = 0$ for the diamagnetic distally ligated O_2 and CO complexes) are restricted to the same subunit as ligand binding. These changes can be used to measure the relative affinities of the α and β chains for different ligands, and to detect structural intermediates at partly saturating ligand concentrations.[19]

A second structural feature of Hb A that has facilitated 1D ^{1}H NMR studies in diamagnetic ligated (O_2 and CO) forms is the large fields near the porphyrin rings of the hemes that results from π electron ring currents. These ring current fields produce large chemical shift perturbations of resonances of many of the hydrogen nuclei in the distal pockets, in particular shifting the distal valine side-chain methyl groups (α62Val and β67Val) to an upfield frequency outside the main bulk of the aliphatic proton resonances. These valine methyl resonances can be used to monitor local structural changes in the distal pockets as a result of changes in conditions or sequence substitution.[19]

A third structural feature of human Hb that has facilitated 1D ^{1}H NMR studies is the formation of stable hydrogen bonds between side chains at the $\alpha_1\beta_1$ and $\alpha_1\beta_2$ interfaces. These hydrogen bonds shift the resonance of the protons involved in the hydrogen bonds downfield, where they can be individually resolved.[19] Because some of these hydrogen bonds (in particular, those of the $\alpha_1\beta_2$ interfaces) are disrupted in either the R or the T quaternary structure, they can be used to assess quaternary structure and how it changes in response to solution conditions and site-specific substitutions.

Although the number of ^{1}H resonances that can be resolved from these three shift mechanisms (hyperfine interaction, ring current effects, and hydrogen bonding) are small compared with the number of hydrogen nuclei in Hb A, this subset of resolved resonances is strategically located within the structure of Hb. This is because the interactions that produce these chemical shift perturbations are located at either the hemes, where ligation and ligand selectivity takes place and where allosteric signals initiate and terminate, or at subunit interfaces, which are the sites of quaternary structure changes. Using these resonances, a great deal has been learned from ^{1}H NMR studies of Hb A; much of this has been reviewed by Ho[19]; more recent findings are presented below.

To obtain a more detailed structural picture of Hb A using NMR spectroscopy, multidimensional and multinuclear methods are needed to resolve resonance overlap that limits interpretation of most of the 1D ^{1}H NMR spectra of Hb A. For homonuclear, multidimensional ^{1}H NMR studies of large proteins, such as Hb A, there are several inherent problems that must be dealt with. In addition to the resolution problem resulting from the large number of protons in Hb A, and the line broadening of ^{1}H resonances that results from slow tumbling, a sensitivity problem results from rapid spin–spin relaxation, which can cause spin coherences to decay during the time required for the transfer of magnetization through bonds. If the ^{1}H linewidth exceeds the typical three-bond coupling $^3J_{HH}$, that is, 3–5 Hz, through-bond correlation spectroscopy (COSY)-type experiments become inefficient. Furthermore, through-space (dipolar) correlation or exchange spectroscopy (nuclear Overhauser effect spectroscopy, NOESY) homonuclear experiments suffer from severe spin diffusion. Thus, one needs the spectral editing power and the more efficient J correlation based on the much larger scalar couplings between ^{1}H nuclei and heteronuclei. In protein NMR spectroscopy, ^{15}N and ^{13}C are most appropriate because they both have spectroscopically favorable nuclear spins of 1/2; however, they are at an impractically low natural abundance (0.37 and 1.1%, respectively). Therefore, an expression system (like the bacterial expression system described below) is needed to produce recombinant Hb A and incorporate isotopes such as ^{13}C, ^{15}N, and ^{2}H into the polypeptide.

An additional advantage of incorporation of ^{13}C and ^{15}N into Hb A is that the chemical shift dispersion of these nuclei is much better than for ^{1}H, permitting highly resolved 2D heteronuclear single- and multiple-quantum coherence spectra (HSQC and HMQC, respectively), in which the chemical shifts of ^{1}H nuclei are correlated with those of directly (and indirectly, for HMQC) bound heteronuclei.[21] As an example of the high-quality 2D spectra of Hb A that can be obtained by taking advantage of these large heteronuclear couplings, a [^{15}N, ^{1}H]-HSQC spectrum of uniformly ^{15}N-labeled rHbCO A is shown in Fig. 2A. This spectrum contains cross-peaks from ^{1}H nuclei directly bonded to ^{15}N nuclei, primarily those of the peptide and side-chain (Gln, Asn) amide bonds. A great deal of spectral overlap that prevents interpretation of the amide region of the 1D ^{1}H NMR spectrum is eliminated, and signals from individual ^{1}H nuclei appear as resolved, ^{15}N-correlated cross-peaks.

[21] J. Cavanagh, W. J. Fairbrother, A. G. Palmer, III, and N. J. Skelton, "Protein NMR Spectroscopy: Principles and Practice." Academic Press, San Diego, CA, 1996.

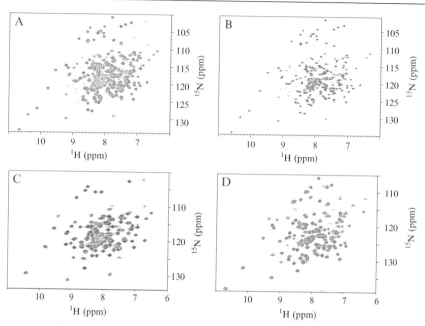

FIG. 2. [^{15}N, ^{1}H] single-bond correlation spectroscopy of uniformly labeled and chain-selective HbCO A. (A) [^{15}N, ^{1}H]-HSQC spectrum of uniformly ^{15}N, ^{2}H-labeled rHbCO A at 600 MHz in 95% H$_2$O–5% D$_2$O. ^{1}H signals result from protons that can exchange with H$_2$O, that is, peptide and side-chain (N)H groups. Although resolution is greatly improved over the analogous 1D ^{1}H NMR spectrum, considerable overlap remains. (B) TROSY spectrum of uniformly ^{15}N, ^{2}H-labeled rHbCO A at 600 MHz, which results in a significant decrease in linewidth, thus decreasing resonance overlap compared with the analogous HSQC spectrum. (C and D) [^{15}N,^{1}H]-HSQC spectra of ^{15}N, ^{2}H, ^{13}C-labeled rHbCO A samples in which labeling is restricted to the α subunits (blue) or the β subunits (red) respectively, thus decreasing overlap compared with the spectrum of the uniformly ^{15}N-labeled sample.[22] (C) and (D) were reproduced from Fig. 2 of Simplaceanu et al.,[22] with permission. (See color insert.)

Although overlap is greatly decreased in 2D [^{15}N,^{1}H] correlation spectra of rHbCO A (Fig. 2A), a great deal of resonance overlap remains, because of the similarity of α and β chains. This overlap can be further reduced by using more recent TROSY techniques (Fig. 2B; and see below), and through the use of chain-selectively labeled samples of Hb A.[22] To prepare chain-selectively labeled samples, uniformly ^{15}N-, ^{2}H-, and/or ^{13}C-labeled Hb A is expressed and purified, and the α and β chains are separated and then recombined with the complementary (β and α, respectively)

[22] V. Simplaceanu, J. A. Lukin, T. Y. Fang, M. Zou, N. T. Ho, and C. Ho, Biophys. J. **79,** 1146 (2000).

unlabeled chains to reform tetramers labeled in either the α or β subunits, but not both. The two separate [^{1}H,^{15}N]-HSQC spectra in Fig. 2C and D correspond to Hb A samples that contain uniformly ^{15}N-labeled α subunits combined with unlabeled β subunits (Fig. 2C), and the reciprocal, uniformly ^{15}N-labeled β subunits combined with unlabeled α subunits (Fig. 2D). In addition to simplifying the spectrum by decreasing the number of resonances by a factor of two, this chain-specific labeling procedure immediately maps each resonance to a particular subunit.[22]

A specific example of the resolution enhancement afforded by multinuclear multidimensional NMR spectroscopy, compared with 1D ^{1}H NMR is shown in Fig. 3, in which an [^{1}H,^{15}N]-HMQC spectrum of fully ^{15}N-labeled HbCO A[22] is compared with a 1D ^{1}H NMR spectrum of HbCO A, published 11 years earlier.[23] In each of these spectra, the portion that highlights histidyl resonances is shown. In addition to providing nitrogen chemical shifts of these histidyl resonances and assignment of the tautomeric state of each histidine,[22] the [^{1}H,^{15}N]-HMQC spectrum is much less congested than the analogous 1D ^{1}H spectrum, resolving the overlap of resonances seen in the corresponding 1D ^{1}H spectrum (compare β116His and β117His resonances with the corresponding resonances labeled I + J + K in the 1D spectrum, Fig. 3).

Three-dimensional NMR can further reduce the spectral complexity of large proteins. One useful 3D NMR experiment is ^{15}N-edited NOESY-HSQC, in which the 2D NOESY spectrum is spread into a third dimension corresponding to the chemical shift of the ^{15}N nucleus bonded to one of the protons. The ^{15}N editing helps resolve the NOESY spectrum, which would be too crowded in just two dimensions. To interpret the NOESY cross-peaks in terms of distance between protons, the ^{1}H and ^{15}N chemical shifts must be assigned to specific nuclei in the protein. A strategy for resonance assignment is to use 3D triple-resonance experiments to sequentially transfer magnetization along specific pathways in an ^{15}N,^{13}C-labeled protein.[24] These experiments provide sets of chemical shift correlations, which collectively cover the polypeptide backbone.

Even multinuclear experiments tend to lose sensitivity in proteins as large as Hb A, because of rapid transverse relaxation. This problem can be somewhat alleviated by ^{2}H labeling, which lengthens ^{13}C, ^{15}N, and H_N relaxation times by reducing the ^{13}C–^{1}H dipolar interactions as well as long-range ^{15}N–^{1}H and ^{1}H–^{1}H dipolar and indirect interactions. A significant improvement in resolution and efficiency has been achieved by the development of transverse relaxation-optimized spectroscopy

[23] I. M. Russu, S.-S. Wu, N. T. Ho, G. W. Kellog, and C. Ho, *Biochemistry* **28**, 5298 (1989).
[24] M. Ikura, L. E. Kay, and A. Bax, *Biochemistry* **29**, 4659 (1990).

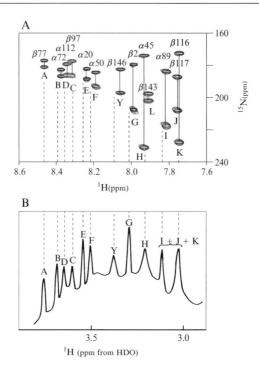

FIG. 3. Enhanced resolution of surface histidyl resonances resulting from multinuclear high-field 2D NMR spectroscopy. (A) A 600-MHz $[^{15}N,^{1}H]$ multiple-bond correlation (HMQC) spectrum of uniformly labeled HbCO A, with resonance assignments listed. Each set of peaks represents a two-bond coupling between a single histidine $^{1}H_{\varepsilon 1}$ and a pair of ^{15}N nuclei ($N_{\varepsilon 2}$ and $N_{\delta 1}$). (B) A 300-MHz 1D ^{1}H NMR spectrum of the same region of HbCO A under otherwise identical conditions. Reproduced from Fig. 5 of Simplaceanu et al.,[22] with permission.

(TROSY).[25] An example of the improvement in resolution afforded by TROSY-based experiments can be seen by comparing the TROSY spectrum of uniformly ^{15}N-labeled HbCO A in Fig. 2B with the analogous HSQC spectrum in Fig. 2A. TROSY-based experiments have been used to assign polypeptide backbone resonances,[26] and to determine the structure and dynamic properties of Hb A in solution.[27,28]

[25] M. Salzmann, K. K. Pervushin, G. Wider, H. Senn, and K. Wüthrich, *Proc. Natl. Acad. Sci. USA* **95,** 13585 (1998).
[26] J. A. Lukin, G. Kontaxis, V. Simplaceanu, A. Bax, and C. Ho, *J. Biomol. NMR.* In press (2003).
[27] Y. Yuan, V. Simplaceanu, J. A. Lukin, and C. Ho, in preparation (2003).
[28] Y. Yuan, V. Simplaceanu, J. A. Lukin, and C. Ho, *J. Mol. Biol.* **321,** 863 (2002).

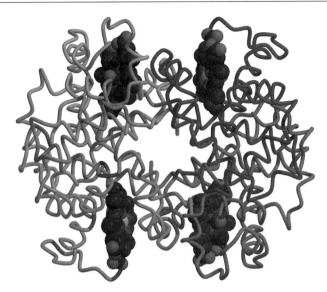

FIG. 4. A comparison of the extent of resonance assignments in HbCO A using ^{1}H NMR *(left)* and by more recent multinuclear [^{13}C,^{15}N] experiments *(right)*. Dark shading (blue for α subunits, red for β subunits) indicates residues for which at least one resonance assignment has been made; light shading indicates residues that are unassigned. The $\alpha_1\beta_1$ dimer on the left shows assignments made before the application of multinuclear multidimensional NMR to isotopically labeled HbCO A, whereas the $\alpha_2\beta_2$ dimer on the right includes assignments made as a result of these methods (Lukin *et al.*[26]; backbone resonances have been deposited at BioMagResBank). Assignment information is superposed on the crystal structure of HbCO A in the R2 form (1BBB[16]). This image was prepared using MOLSCRIPT[4] and Raster3D.[5] (See color insert.)

For Hb A, the ability to incorporate isotopes such as ^{15}N and ^{13}C has facilitated a major advance in resonance assignments. This can be seen in Fig. 4, where amino acid residues in HbCO A for which no assignments have been made are in lightly shaded colors (blue for the α chain and red for the β chain), whereas residues for which at least one assignment has been made are shaded with full saturation. The $\alpha_1\beta_1$ dimer on the left side of Fig. 4 shows HbCO A assignments before the application of multinuclear NMR to isotopically labeled rHb A,[19,29,30] whereas the $\alpha_2\beta_2$ dimer on the right shows the current level of assignment, facilitated by application of multinuclear NMR to isotopically labeled rHbCO A (Lukin *et al.*[26]; backbone resonances have been deposited at BioMagResBank, accession

[29] D. P. Sun, M. Zou, N. T. Ho, and C. Ho, *Biochemistry* **36,** 6663 (1997).
[30] T. Y. Fang, M. Zou, V. Simplaceanu, N. T. Ho, and C. Ho, *Biochemistry* **38,** 13423 (1999).

number 5856). As can be seen in Fig. 4, the application of multinuclear NMR to isotopically labeled rHbCO A permitted nearly full coverage of assignments, especially in the β subunits. On average, 4.5 separate assignments (^1H, ^{15}N, and ^{13}C) have been made at each residue for which at least one assignment is available (dark shading, Fig. 4). This high level of assignments provides extensive probes to monitor distal ligand binding and allosteric change, and will facilitate the determination of a high-resolution solution structure using inter proton distances (from $^1H-^1H$ NOE measurements) and chemical shift values.

Determination of high-resolution protein structure by NMR spectroscopy relies primarily on short-range interactions (less than 6 Å for the NOE, within vicinal covalently bonded systems for ^3J). For small, monomeric proteins, these short-range interactions can sufficiently determine global protein structure to a high degree of accuracy. However, for large, multimeric proteins such as Hb A, the lack of long-range structural probes can lead to considerable uncertainty regarding the structural relationship between distant parts of the protein, such as relative orientation of individual subunits. One way to obtain long-range structural information by NMR spectroscopy is to impart a partial alignment on the protein. If a small degree of alignment is imposed on a protein by a medium such as phospholipid bicelles or filamentous phage (Pf1), each amide ^1H, ^{15}N pair will experience a residual dipolar coupling that would add to or subtract from the isotropic one-bond J coupling.[31] The magnitude of this residual dipolar coupling depends on the orientation of the N–H bond vectors with respect to the overall axis of alignment of the protein. Since for a rigid protein molecule, this axis is the same for all parts of the molecule, it produces a means to connect structural features in all parts of the protein. As discussed below, residual dipolar coupling methods have been used to examine the quaternary structure of HbCO A in solution.[32]

Artificial Means of Production of Hb A

To prepare isotopically labeled Hb A for multinuclear NMR experiments, and to introduce precise sequence substitutions, a system is needed that can produce correctly assembled tetrameric Hb A in a simple expression system such as a bacterial culture. One approach to this problem is to use a bacterial, high-copy plasmid to direct the synthesis of both the α and β chains within the same bacterial cell.[33,34] This approach has the benefit

[31] E. de Alba and N. Tjandra, *Prog. NMR Spectrosc.* **40,** 175 (2002).
[32] J. A. Lukin, G. Kontaxis, V. Simplaceanu, Y. Yuan, A. Bax, and C. Ho, *Proc. Natl. Acad. Sci. USA* **100,** 517 (2003).

that, in principle, it should lead to stoichiometric production of α and β subunits, maximizing the formation of $\alpha_2\beta_2$ heterotetramers. Using a bicistronic plasmid in which each globin gene is under the control of the same promoter, similar levels of expression of the α and β chains have been obtained.[34] Another advantage of expressing and assembling the α and β chains in the same cell is that *in vitro* reconstitution of purified proteins[35,36] is avoided.

One of the drawbacks of a bacterial expression system is that polypeptides synthesized in bacteria begin with an amino-terminal methionyl residue. For many monomeric proteins, the presence of this additional methionyl residue can be overlooked, because the termini are usually surface exposed in the folded, native state. However, in Hb A the N terminus makes a number of intersubunit contacts,[9,37] and the addition of an N-terminal methionyl residue is likely to perturb tertiary and quaternary structure and, thus, reactivity. This problem has been resolved by coexpressing methionine aminopeptidase at high levels along with the α and β chains of Hb A.[34,38] Methionine aminopeptidase acts to remove N-terminal methionyl residues, leaving the second residue as the new amino terminus. Using mass spectrometry, this method has been shown to produce correctly processed α and β chains with the same sequence as the chains of native Hb A.[34,38]

As an alternative strategy, α and β chains can be expressed that are missing the first (amino-terminal) residue (a valine in both subunits).[39] Although this strategy produces α and β chains of the same length as in native Hb A, it results in sequence substitution of each of the first residues in the four subunits (Val→Met). Whereas $\alpha_2\beta_2$ tetramers produced by this method are quite similar to native human Hb A, these N-terminal sequence substitutions do produce subtle differences in reactivity and structure.[39,40] Other expression systems, including yeast[41] and transgenic

[33] S. J. Hoffman, D. L. Looker, J. M. Roehrich, P. E. Cozart, S. L. Durfee, J. L. Tedesco, and G. L. Stetler, *Proc. Natl. Acad. Sci. USA* **87,** 8521 (1990).

[34] T. J. Shen, N. T. Ho, V. Simplaceanu, M. Zou, B. N. Green, M. F. Tam, and C. Ho, *Proc. Natl. Acad. Sci. USA* **90,** 8108 (1993).

[35] K. Nagai and H. C. Thogersen, *Nature* **309,** 810 (1984).

[36] K. Nagai, M. F. Perutz, and C. Poyart, *Proc. Natl. Acad. Sci. USA* **82,** 7252 (1985).

[37] R. E. Dickerson and I. Geis, "Hemoglobin." Benjamin/Cummings, Menlo Park, CA, 1983.

[38] T. J. Shen, N. T. Ho, M. Zou, D. P. Sun, P. F. Cottam, V. Simplaceanu, M. F. Tam, D. A. Bell, Jr., and C. Ho, *Protein Eng.* **10,** 1085 (1997).

[39] H. L. Hui, J. S. Kavanaugh, M. L. Doyle, A. Wierzba, P. H. Rogers, A. Arnone, J. M. Holt, G. K. Ackers, and R. W. Noble, *Biochemistry* **38,** 1040 (1999).

[40] J. S. Kavanaugh, P. H. Rogers, and A. Arnone, *Biochemistry* **31,** 8640 (1992).

[41] M. Wagenbach, K. O'Rourke, L. Vitez, A. Wieczorek, S. Hoffman, S. Durfee, J. Tedesco, and G. Stetler, *BioTechnology* **9,** 57 (1991).

swine,[42] have been reported to produce $\alpha_2\beta_2$ tetramers of Hb A. Although the swine system has the advantage that large quantities of Hb are produced that are functionally and structurally identical to Hb A,[42] the use of this system for isotopic labeling would be impractical at best.

An additional challenge in producing native Hb A from *Escherichia coli* arises from the fact that the Hb tetramer is a large noncovalent assembly, and that there are many opportunities for these noncovalent interactions to form in nonnative ways. Specifically, α- and β-globin chains must combine properly with a heme, and each of these assembled subunits must combine to form an $\alpha_2\beta_2$ heterotetramer with native intersubunit interactions. Whereas obtaining the correct intersubunit interactions between α and β chains is likely to be thermodynamically controlled (and thus the correct intersubunit interactions are likely to form spontaneously), proper insertion of heme into globin chains can fall under kinetic control: for myoglobin, heme binds to the apoprotein in more than one conformation, and interconverts on a time scale of hours to weeks, depending on the oxidation state and nature of the distal ligand.[43-45] Indeed, recombinant Hb A tetramers produced in *E. coli* show significant heterogeneity.[34,38,46] Fortunately, this heterogeneity can be eliminated by oxidation of the heme iron, which presumably facilitates heme isomerization, followed by reduction to the ferrous state[34,38]: the product of this redox cycle appears to be identical to native human Hb A. Interestingly, the degree to which heme is correctly inserted varies with the source of expression. For instance, Hb A synthesized in yeast shows more heme disorder than when synthesized with a rabbit reticulocyte lysate,[47] although the Hb A tetramers produced from both systems showed significant heterogeneity compared with native Hb A derived from human erythrocytes. This observation suggests that there is a specific mechanism or pathway by which heme is correctly inserted in erythrocytes, a suggestion consistent with the observation that Hb A isolated from the erythrocytes of transgenic swine have the same low level of heme disorder as native Hb A from human erythrocytes.[42]

Another benefit of producing rHb A in a heterologous expression system is that site-specific sequence substitutions can be introduced into any site within the α and β subunits, and the effects of precise sequence

[42] B. N. Manjula, R. Kumar, D. P. Sun, N. T. Ho, C. Ho, J. M. Rao, A. Malavalli, and A. S. Acharya, *Protein Eng.* **11**, 583 (1998).
[43] T. Jue, R. Krishnamoorthi, and G. N. La Mar, *J. Am. Chem. Soc.* **105**, 5701 (1983).
[44] G. N. La Mar and R. Krishnamoorthi, *J. Am. Chem. Soc.* **106**, 6395 (1984).
[45] J. T. J. Lecomte, R. D. Johnson, and G. N. La Mar, *Biochim. Biophys. Acta* **829**, 268 (1985).
[46] R. A. Hernan and S. G. Sligar, *J. Biol. Chem.* **270**, 26257 (1995).
[47] A. J. Mathews and T. Brittain, *Biochem. J.* **357**, 305 (2001).

substitution on distal ligand reactivity, structure, and allosteric switching can be determined. Using a variety of site-directed mutagenesis methods to engineer mutations to the genes encoding the Hb A chains (for a review of methods to introduce site-specific substitutions, see Barrick[48]), sequence substitutions have been engineered at the proximal and distal heme pockets, and at the subunit interfaces (Fig. 5). ^{1}H NMR spectroscopy is an excellent tool to rapidly determine the structural and functional consequences of sequence substitutions. In the following sections, several examples of combining NMR studies with sequence-substituted rHbs are described; refer to Fig. 5 to visualize the locations of, and interactions made by, substituted amino acid residues described below.

Use of NMR Spectroscopy to Examine Structure and Reactivity at
  Heme Sites

Proximal Pocket and Allosteric Mechanism

The fact that the oxygen-binding sites in Hb A coincide with sites of large, local magnetic fields (from heme ring currents and paramagnetic centers) makes NMR an ideal tool to monitor distal ligand binding, determine the structural features of the ligand-binding sites, and identify local structural changes that accompany the binding reaction. Understanding these structural changes is important for determining the distal ligand affinity and specificity, and for understanding the initial and final stages of allosteric interaction between the heme sites.

An example of the use of 1D ^{1}H NMR to examine ligand affinity and allostery comes from the study of the hyperfine-shifted proximal ligand resonances in proximally detached hemoglobins.[13,14] As outlined above, proximal detachment is a method that was used to substitute the proximal histidine ligands *in trans* with imidazoles, which mimic the local bonding interactions of the proximal histidine side chains, but lack covalent linkage with the F helices, thus testing the Perutz model for allostery. Because in deoxy-Hb A, the heme irons are high-spin ($S = 2$), nearby protons experience large chemical shift perturbations. Two resonances in the extreme downfield region (63.3 and 75.8 ppm) of deoxy-Hb A have been assigned to the (N_δ)Hs of the proximal histidines of the α and β subunits, respectively.[49–51] Owing to the high sensitivity of the chemical shifts of these

[48] D. Barrick, *in* "Comprehensive Coordination Chemistry II" (A. B. P. Lever, ed.), Vol. 1. Pergamon Press, Oxford, 2003.
[49] S. Takahashi, A. K. Lin, and C. Ho, *Biochemistry* **19**, 5196 (1980).
[50] G. N. La Mar, D. L. Budd, and H. Goff, *Biochem. Biophys. Res. Commun.* **77**, 104 (1977).
[51] G. N. La Mar, U. Kolczak, A. T. Tran, and E. Y. Chien, *J. Am. Chem. Soc.* **123**, 4266 (2001).

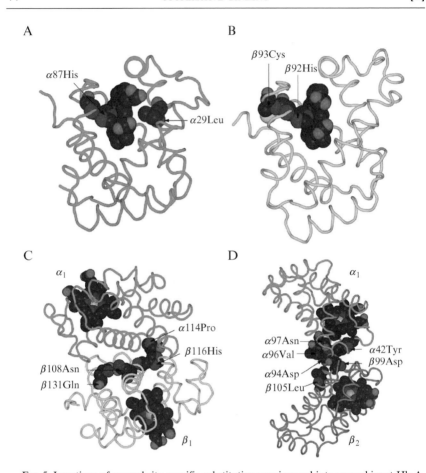

FIG. 5. Locations of several site-specific substitutions engineered into recombinant Hb A. (A and B) Residues substituted in the heme pockets of the α subunit (yellow) and β subunit (green), respectively, with heme groups represented by van der Waals spheres. (C) Residue substitutions at the $\alpha_1\beta_1$ interface (α and β subunits in green and yellow, respectively). Residue substitutions at the $\alpha_1\beta_2$ interfaces (α and β subunits in green and orange, respectively). This image was prepared using MOLSCRIPT[4] and Raster3D.[5] (See color insert.)

protons to their distance from, and details of bonding with the heme iron, these resonances serve as excellent probes of structural features of the proximal pocket. On binding of distal ligand, the heme irons become diamagnetic ($S = 0$) and these resonances disappear from the spectrum. Thus, these two resonances also serve as excellent subunit-specific probes of distal ligand binding.

In the proximally detached rHb (αH87G) variant, in which the linkages between the α-subunit proximal ligands and the polypeptide backbone are severed, an overall T-quaternary structure appears to be strongly favored both in the absence and in the presence of distal ligand binding.[13] Thus, the downfield region of the ^{1}H NMR spectrum of deoxy-rHb (αH87G) should show an unperturbed β-subunit (N$_\delta$)H resonance at 75.8 ppm, but the resonance associated with the imino (N)H of the α-subunit proximal imidazoles should be substantially perturbed compared with the 63.3-ppm resonance of the analogous α-subunit proximal histidines of wild-type deoxy-Hb A. Consistent with these predictions, the downfield region of the deoxy-Hb (αH87G) shows a resonance at 74.5 ppm, close to the β-subunit proximal histidine resonance of deoxy-Hb A, but lacks a resonance at the frequency corresponding to the α-subunit proximal histidine resonance of deoxy-Hb A.[14] Instead, two resonances are seen at 54 and 82 ppm, presumably from the (C$_\gamma$)H and (N$_\delta$)H protons. By preparing analogous deoxy-Hb (αH87G) samples in deuterium oxide (D$_2$O) and using deuterated d_4-imidazole, thus selectively deuterating these two positions, the 54- and 82-ppm resonances could be assigned to the α-subunit imidazole (C$_\gamma$)H and (N$_\delta$)H protons, respectively.[14]

With subunit-specific proximal assignments in hand, the sensitivity of the proximal resonances to distal ligand binding could be used to dissect the distal ligand-binding curve into its site-specific components. Resolving the distal ligand-binding curve into subunit-specific interactions was particularly informative for the proximally detached rHb (αH87G), as the overall distal ligand-binding curve of this protein yielded a Hill coefficient of about 0.8.[13] A Hill coefficient of less than one is compatible with either binding-site heterogeneity, or with anticooperativity. By titrating deoxy-Hb (αH87G) with n-butyl isocyanide (n-BuNC, a distal ligand that binds to wild-type Hb A with positive cooperativity) and collecting ^{1}H NMR spectra in the downfield region, the broad overall distal ligand-binding curve could be attributed to high-affinity α subunits and low-affinity β subunits within the tetramer.[14] This observation is consistent with the hypothesis that the proximal linkage plays an important role in communicating allosteric interactions between the heme sites, and in limiting reactivity in the T state. Because rHb (αH87G) remains in the T-quaternary structure on distal ligation, the affinity of the β subunits remains low because the β-subunit proximal linkages remain intact. In contrast, the α subunits have high affinity despite the overall T-quaternary structure because they are freed from proximal restraint. The ability of NMR spectroscopy to provide baseline-resolved, subunit-specific probes in solution was critical for understanding the thermodynamic basis for distal ligand binding in this and other proximally detached Hbs, and thus for testing the Perutz model for allostery. Studies of these proximally detached rHbs[13,14] led to the following conclusions: (1) the proximal coupling mechanism as proposed by Perutz contributes approximately two-thirds of the total interaction energy between the hemes, the remaining third resulting from alternate coupling pathways; (2) in distally ligated Hb, the T state is compatible with proximal linkages in the β chains but not the α chains; and (3) comparing ligand association and dissociation rates between rHb (αH87G) with the T and R states of Hb A indicates that at the α chains, CO affinity is modulated entirely by the proximal linkage, rather than from distal interactions; some residual allosteric interactions may remain operative at the β chains of rHb (αH87G).

Distal Pocket, Ligand Orientation, and Hydrogen Bonding

The structural basis of distal ligand affinity in heme proteins has been the focus of intense research. One area of interest has centered on the affinity difference between carbon monoxide and oxygen (reviewed by Spiro and Kozlowski[52]). Although CO binds more tightly than O_2 to respiratory heme proteins such as myoglobin and hemoglobin, by between 25- and 250-fold,[1] CO binds more tightly than O_2 to heme model compounds, by about 20,000-fold.[53] The modulation of the relative affinity of these two ligands that results from encapsulation of the heme within the protein matrix of the globin is critical for function, since a high affinity for CO relative to O_2 in respiratory heme proteins would result in CO saturation at levels of CO normally found in the circulatory system, which in turn would lead to asphyxiation.[54]

One model by which the protein is proposed to diminish the affinity of CO relative to that of O_2 is by unfavorable steric interaction with the bound CO ligand.[53] Such a differential effect is suggested by the observation that CO and O_2 bind differently to simple heme model compounds: CO binds in a straight-on, linear fashion, whereas O_2 binds with a bent geometry. Early crystal structures suggested that for myoglobin, CO is bound in a bent geometry,[55] consistent with the proposal that the CO ligand is sterically prevented from binding in its most favored (linear) orientation by close contacts with the protein matrix, thus weakening its interaction.[53] However, subsequent studies have cast doubt on this steric mechanism, partly because a large number of high-resolution crystal structures have suggested both that

[52] T. G. Spiro and P. M. Kozlowski, *Acc. Chem. Res.* **34,** 137 (2001).
[53] J. P. Collman, J. I. Brauman, T. R. Halbert, and K. S. Suslick, *Proc. Natl. Acad. Sci. USA* **73,** 3333 (1976).
[54] B. A. Springer, S. G. Sligar, J. S. Olson, and G. N. Phillips, Jr., *Chem. Rev.* **94,** 699 (1994).
[55] J. Kuriyan, S. Wilz, M. Karplus, and G. A. Petsko, *J. Mol. Biol.* **192,** 133 (1986).

a variety of less bent structures can be accommodated, and partly because the subtle variation in electron density that would be produced by minor variations in ligand position may not be discernible even in structures of high resolution and quality.[52]

Another model by which the protein is proposed to diminish the affinity of CO relative to that of O_2 invokes the formation of a stabilizing hydrogen bond between the distal histidine $N_\varepsilon H$ and the distal O_2 ligand, increasing the affinity of O_2 relative to the CO ligand. Although this model is based on structural features of the distal pocket of heme proteins determined by X-ray crystallography, the structural details necessary to unequivocally confirm the presence of these proposed hydrogen bonds are not easy to determine from crystal structures. Two factors that contribute to this are the inability of X-ray crystallography to determine the tautomeric state of the distal histidine (critical to evaluate the ability of the histidine to donate a hydrogen bond) and again the precise orientation of the distal ligand.

Both the detailed orientation of distal O_2 and CO ligands, as well as the details of hydrogen bonding, can be investigated by NMR spectroscopy. By collecting [^{15}N, ^{1}H]-HMQC spectra on ^{15}N-labeled HbO_2 A and HbCO A, which detects and discriminates both single- and multiple-bond coupling between ^{1}H and ^{15}N nuclei, allowing the tautomerization state of histidines to be determined, the distal histidines of both the α and β subunits of HbO_2 A and HbCO A were shown unequivocally to be in the neutral $N_\varepsilon H$ tautomeric form, consistent with hydrogen bonding to the distal histidines and both distal O_2 and CO ligands.[56] Further, these spectra show that in HbO_2 A, but not in HbCO A, the distal histidine $N_\varepsilon H$ hydrogens are protected from solvent exchange, consistent with stabilizing hydrogen bond formation in HbO_2 A, relative to that seen in HbCO A.[56] Although the increased rate of exchange of the distal histidine $N_\varepsilon H$ hydrogen in HbCO A indicates that any distal hydrogen bonding is weak relative to that in HbO_2 A, it does not imply that hydrogen bonding is absent between the distal histidine and the distal CO ligand. By comparing measured ^{1}H chemical shifts with those calculated from various HbCO and HbO_2 crystal structures, using a program that incorporates ring current, magnetic anisotropy, and electric field effects,[57] structural differences could be identified between the distal pockets of the α and β subunits of HbO_2 in solution,[56] and between both the α and β subunits of HbO_2 A in solution and the crystal structure.[58]

[56] J. A. Lukin, V. Simplaceanu, M. Zou, N. T. Ho, and C. Ho, *Proc. Natl. Acad. Sci. USA* **97,** 10354 (2000).

[57] M. P. Williamson and T. Asakura, *J. Magn. Reson. Ser. B* **101,** 63 (1993).

[58] B. Shannon, *J. Mol. Biol.* **171,** 31 (1983).

The structure of the distal pocket and the presence of hydrogen bonding between the distal histidines and the distal ligand have also been examined in the cyanomet complex of ferric Hb A (HbCN A), using ^{1}H NMR spectroscopy.[51] The CN$^-$ ligand has the capacity for both steric and hydrogen bonding interactions with the distal protein matrix,[51] and the cyanomet form of Hb A is considered to provide a reasonable model of the R-state hemoglobin. Several spectroscopic observations support the presence of hydrogen bonds between the distal histidines of both the α and β subunits and the bound CN$^-$ ligand. First, as with the HbO$_2$ A complex, the N$_\varepsilon$H in the α subunit shows significant protection from exchange with solvent.[51] Second, using hyperfine shift and relaxation data, the distal histidines in the α subunits were determined to be repositioned towards the distal CN$^-$, in an orientation consistent with hydrogen bonding. Although these observations could not be extended to the distal histidines of the β subunits, the average contact shift of the heme methyls, which has been shown to correlate with hydrogen bonding to the distal CN$^-$ ligand,[59] was found to be similar between the α and β subunits, consistent with a hydrogen bond to the distal CN$^-$ ligand in both subunits.[51]

Both of these studies indicate that hydrogen bonding is present between distally bound ligands and the distal histidines of the α and β subunits, and demonstrate how NMR can be used to identify subtle but functionally important structural differences between the solution and crystalline states.[51,56] The observation that the distal histidine N$_\varepsilon$Hs are protected in HbO$_2$ A, but not in HbCO A, is consistent with distal hydrogen bonding as a means to decrease the affinity difference between the two ligands. This conclusion is consistent with extensive mutagenesis studies in myoglobin,[60] which show that substitution of the distal histidine with bulky non-hydrogen-bonding residues sharply decreases affinity for O$_2$ but not for CO.

Use of NMR Spectroscopy to Investigate Involvement of Subunit Interfaces in Hemoglobin Allostery

Persistent emphasis has been placed on the role of changes in quaternary structure on hemoglobin allostery. As described above, much of the focus has been placed on the conformational changes at the $\alpha_1\beta_2$ (and the symmetry-related $\alpha_2\beta_1$) interfaces that occur on distal ligand binding

[59] B. D. Nguyen, Z. Xia, F. Cutruzzola, C. Travaglini Allocatelli, A. Brancaccio, M. Brunori, and G. N. La Mar, *J. Biol. Chem.* **275**, 742 (2000).

[60] J. S. Olson and G. N. Phillips, Jr., *J. Biol. Inorg. Chem.* **2**, 544 (1997).

(i.e., between $\alpha\beta$ dimers), as these changes are much larger than those at $\alpha_1\beta_1$ (and the symmetry-related $\alpha_2\beta_2$) interfaces (i.e., within $\alpha\beta$ dimers).[12]

NMR spectroscopy has been used as a tool to dissect the role of specific interactions within the $\alpha_1\beta_2$ and $\alpha_2\beta_1$ interfaces in quaternary structure switching. Substitution of several residues that make intersubunit interactions at these interfaces in deoxy-Hb A have been shown to decrease the stability of the T-quaternary structure, favoring the R-quaternary structure, as monitored by the appearance and disappearance of key structural markers in the exchangeable region of the ^{1}H NMR spectrum. Such substitutions tend to increase oxygen affinity and decrease cooperativity. Examples include substitution with Asn of β99Asp, which acts as an intersubunit ($\alpha_1\beta_2$) hydrogen bond donor to α42Tyr and α97Asn.[61] Similar results were obtained by substituting the corresponding hydrogen bond acceptors.[62–64] By simultaneously substituting β99Asp with Asn and α42Tyr with Asp, thus reversing, but preserving the hydrogen bond, T-state quaternary structure markers were found to reappear in the ^{1}H NMR spectrum, oxygen affinity was found to decrease, and cooperativity was partially restored compared with the single βD99N substitution,[61] indicating that the stability imparted by those residues to the T-quaternary structure results from a mutual, pairwise effect, as would be expected for a stabilizing hydrogen bond. Similarly, by substituting α42Tyr with histidine, a residue capable of donating a hydrogen bond to β99Asp, the T-state quaternary structure was restored in the deoxy form compared with a substitution of α42Tyr with phenylalanine, which cannot form a hydrogen bond.[62,63]

Conversely, several studies have introduced interactions into the $\alpha_1\beta_2$ interface that are designed to stabilize the T-quaternary structure, thereby lowering oxygen affinity without sacrificing cooperativity in distal ligand binding. For example, by making a substitution (α96Val→Trp) that creates an intersubunit hydrogen bond at the $\alpha_1\beta_2$ interface in the T state,[65] oxygen affinity could be decreased, while maintaining high cooperativity.[66] Interestingly, for this substitution, ^{1}H NMR studies indicate that the T structure

[61] H. W. Kim, T. J. Shen, D. P. Sun, N. T. Ho, M. Madrid, M. F. Tam, M. Zou, P. F. Cottam, and C. Ho, *Proc. Natl. Acad. Sci. USA* **91**, 11547 (1994).

[62] K. Imai, K. Fushitani, G. Miyazaki, K. Ishimori, T. Kitagawa, Y. Wada, H. Morimoto, I. Morishima, D. T. Shih, and J. Tame, *J. Mol. Biol.* **218**, 769 (1991).

[63] K. Ishimori, I. Morishima, K. Imai, K. Fushitani, G. Miyazaki, D. Shih, J. Tame, J. Pegnier, and K. Nigai, *J. Biol. Chem.* **264**, 14624 (1989).

[64] H. W. Kim, T. J. Shen, N. T. Ho, M. Zou, M. F. Tam, and C. Ho, *Biochemistry* **35**, 6620 (1996).

[65] Y. A. Puius, M. Zou, N. T. Ho, C. Ho, and S. C. Almo, *Biochemistry* **37**, 9258 (1998).

[66] H. W. Kim, T. J. Shen, D. P. Sun, N. T. Ho, M. Madrid, and C. Ho, *J. Mol. Biol.* **248**, 867 (1995).

can be populated in the distally ligated (CO) form by addition of IHP or by lowering the temperature, consistent with a stabilization of the T-quaternary structure by the new hydrogen bond.[66] Similarly, by substituting β105Leu with Trp, with the goal of creating an intersubunit $\alpha_1\beta_2$ hydrogen bond between the substituting Trp indole NH and α94Asp, the oxygen affinity was lowered, but binding remained cooperative.[67] Although the designed hydrogen bond has not been verified crystallographically, [^{15}N,^{1}H]-NMR spectroscopy indicates that the indole NH of the substituting tryptophan is indeed involved in a hydrogen bond in the T-quaternary structure.[67]

Insight into the interactions and dynamics of the $\alpha_1\beta_2$ subunit interface has been obtained by a variety of NMR methods. Two studies have focused on a tryptophan (β37Trp) that is located at a "flexible joint" at the $\alpha_1\beta_2$ interface.[28,68] Hydrogen–deuterium exchange measurements reveal that the indole (N)H of βTrp37 is significantly more labile in the R state than in the T state, suggesting that the contacts made by β37Trp are significantly weaker in the R state than in the T state.[68] Measurements of dynamic properties of the β37Trp indole NH bond vector, using ^{15}N-labeled wild-type and variant Hbs, suggest that the side chain of β37Trp undergoes significant conformational exchange in "strained" HbCO complexes, such as HbCO A in the presence of IHP.[28] Both of these studies indicate that the "flexible joint" region of the $\alpha_1\beta_2$ interface is involved in the allosteric mechanism, and suggest that dynamic features of the interface are linked to the cooperativity.

Although the $\alpha_1\beta_1$ and symmetry-related $\alpha_2\beta_2$ interfaces undergo comparatively small structural changes in the Hb A quaternary structure transition, several studies have suggested that interactions within this interface are coupled to distal ligand binding, and thus may contribute to allostery. In a study of the $N_{\varepsilon 2}$H protons of two histidines (α103His and α122His) that form intersubunit hydrogen bonds at the $\alpha_1\beta_1$ and $\alpha_2\beta_2$ interfaces, hydrogen exchange rates with solvent were found to be significantly faster in the T state than in the R state, suggesting that the strength of these hydrogen bonds changes as a result of the quaternary structure transition,[69] increasing in strength in the R-quaternary structure. Similar conclusions were reached in a mutagenesis study that targeted the intersubunit hydrogen bond acceptor of one of these $\alpha_1\beta_1$ histidines (β131Gln, which hydrogen bonds with the $N_{\varepsilon 2}$H of α103His).[70] Substitution of β131Gln

[67] T. Y. Fang, V. Simplaceanu, C. H. Tsai, N. T. Ho, and C. Ho, *Biochemistry* **39,** 13708 (2000).

[68] M. R. Mihailescu, C. Fronticelli, and I. M. Russu, *Proteins* **44,** 73 (2001).

[69] M. R. Mihailescu and I. M. Russu, *Proc. Natl. Acad. Sci. USA* **98,** 3773 (2001).

[70] C. K. Chang, V. Simplaceanu, and C. Ho, *Biochemistry* **41,** 5644 (2002).

affects the chemical environment of the $N_{\varepsilon2}Hs$ of both $\alpha103His$ and $\alpha122His$, increases the exchange rates of both $N_{\varepsilon2}Hs$ in the distally ligated form, and decreases oxygen affinity.[70] These substitutions also increase the rate of chemical modification of $\beta93Cys$, a residue near the proximal pocket that makes contacts in the $\alpha_1\beta_2$ interface in the distally ligated form, suggesting that these $\alpha_1\beta_1$ intersubunit interactions may be coupled to the $\alpha_1\beta_2$ interface, as well as to distal ligand binding.[70]

The sensitivity of resonances at the $\alpha_1\beta_1$ interfaces to distal ligand binding is consistent with work from the Ackers laboratory, which shows that asymmetric substitutions that target only one of the two interfaces that undergo major rearrangement (e.g., $\alpha_1\beta_2$ but not the symmetry-related $\alpha_2\beta_1$) perturb distal ligand binding only in the dimer in which the substitution is made.[71] Patterns of cooperative free energy penalties in this asymmetrically substituted system, as well as in unsubstituted Hb A,[72] suggest that allosteric interactions are indeed transmitted between α and β subunits of the same $\alpha\beta$ dimer, that is, across the $\alpha_1\beta_1$ and the $\alpha_2\beta_2$ interfaces. Although these cooperative free energy patterns and dimer-autonomous effects are at odds with an O_2-binding study to zinc-substituted Hbs,[73] they are consistent with the NMR and mutational studies described above in suggesting that, although to a first approximation the $\alpha_1\beta_1$ and $\alpha_2\beta_2$ interfaces are structurally invariant as determined crystallographically, the strength and details of intersubunit interactions within these interfaces are coupled to distal ligand binding and to the allosteric transition in Hb A.

Use of NMR Spectroscopy to Probe Quaternary Structure of Hb A in Solution

In addition to probing quaternary structure through the observation of T- and R-state quaternary structure markers in 1D ^1H NMR spectra, NMR spectroscopy has been used to obtain high-resolution structural information about the quaternary structure of Hb A.[32] By orienting ^{15}N-labeled rHbCO, using bicelles[74] and filamentous Pf1 phage,[75] and measuring residual dipolar couplings ($^1D_{NH}$), various quaternary structure arrangements (such as the crystallographically observed R and R2 structures, as well as structures intermediate between these limits) could be evaluated

[71] G. K. Ackers, P. M. Dalessio, G. H. Lew, M. A. Daugherty, and J. M. Holt, *Proc. Natl. Acad. Sci. USA* **99**, 9777 (2002).
[72] G. K. Ackers, *Adv. Protein Chem.* **51**, 185 (1998).
[73] K.-M. Yun, H. Morimoto, and N. Shibayama, *J. Biol. Chem.* **277**, 1878 (2002).
[74] C. R. Sanders, II and J. P. Schwonek, *Biochemistry* **31**, 8898 (1992).
[75] M. R. Hansen, L. Mueller, and A. Pardi, *Nat. Struct. Biol.* **5**, 1065 (1998).

by comparing $^1D_{NH}$ values predicted from these structures with those determined experimentally.[32] Calculated $^1D_{NH}$ values from tetrameric R and R2 crystal structures show significantly larger deviations from measured values than $^1D_{NH}$ values calculated from $\alpha_1\beta_1$ dimers, indicating that although the largely invariant $\alpha_1\beta_1$ dimer in crystals matches the $\alpha_1\beta_1$ solution structure, neither of the crystallographically observed quaternary structures represents the dominant structure of HbCO A in solution. Better agreement was obtained with a hypothetical structure midway between the R and R2 quaternary structures, generated either by least-squares fit of the relative orientation of the $\alpha_1\beta_1$ and $\alpha_2\beta_2$ dimers along the R→R2 pathway, or through a rigid-body molecular dynamics method. This result suggests either that in solution, HbCO A adopts a distinct, static structure midway between R and R2, or that observed $^1D_{NH}$ values result from a dynamic average of roughly equal populations of R and R2 structures[32] (Fig. 6). ^{15}N relaxation rates of backbone amides at the $\alpha_1\beta_2$ interface indicate the presence of significant exchange between different

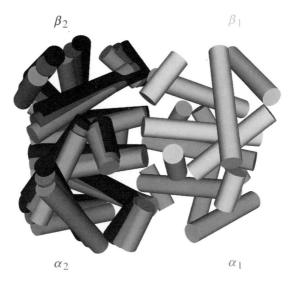

FIG. 6. The quaternary structure of HbCO A in solution, compared with the R and R2 crystal structures. The $\alpha_1\beta_1$ dimers (green and yellow, *right*) of each of the three structures were superposed, facilitating comparison of the relative orientations of the $\alpha_2\beta_2$ dimers (red and blue, *left*). For the high-salt R-state crystal structure 1IRD, the $\alpha_2\beta_2$ dimer is shown with dark shading; for the low-salt R2 crystal structure (1BBB[16]), the $\alpha_2\beta_2$ dimer is shown with light shading; and for the solution structure determined using residual dipolar couplings,[32] the $\alpha_2\beta_2$ dimer is shown with intermediate shading. Reproduced from Fig. 2 of Lukin *et al.*,[32] with permission. (See color insert.)

conformations, consistent with the interpretation that distally ligated Hb A exists as a dynamically interconverting population of different quaternary structures, and that the process of crystallization selects one of these forms based on biases imposed by solution conditions used to promote crystal growth and by intermolecular interactions within the growing crystal lattice.[32]

The observation that multiple ligated quaternary structures are populated in solution may have important functional consequences. If the affinities of the R and R2 structures for distal ligands differ, the relative populations of the R and R2 states would necessarily change with the number of distal ligands bound. Given that the two quaternary structures have similar populations in the fully ligated form,[32] at lower levels of distal ligation, the lower affinity quaternary structure would predominate. This shift would result in positive cooperativity in distal ligand binding within the manifold of R-state structures, consistent with studies of assembly free energies of partly ligated Hbs, which indicate that the fourth distal ligand binds to Hb A with a higher affinity than the third ligand.[72]

Summary and Future Directions

Bacterial expression systems have greatly advanced NMR-based studies of Hb A and our understanding of the relationship between structure and function in Hb. These advances result largely from the ability to isotopically label HbA for high-resolution heteronuclear NMR studies, literally opening up new dimensions in the NMR of hemoglobin, permitting extensive resonance assignments to be made, providing information about dynamics and hydrogen bonding, and yielding precise information regarding quaternary structure in solution. These expression systems also permit site-specific substitutions to be introduced to the Hb tetramer. NMR spectroscopy provides a powerful tool for quickly determining the effects of site-specific substitutions on the structure of the heme pocket, on site-specific distal ligand binding, on subunit interface structure and dynamics, and on the overall quaternary structure in solution.

One area where NMR spectroscopy is likely to further advance our understanding of hemoglobin allostery is conformational dynamics. As described above, a few sites at the $\alpha_1\beta_2$ subunit interface appear to be quite dynamic over a range of time scales, and these dynamics appear to be coupled to distal ligand binding and quaternary structure. By fully characterizing the range of dynamics of different forms of Hb, NMR spectroscopy will provide a picture of hemoglobin allostery that is considerably more detailed than that provided by the static structures obtained from X-ray crystallography.

Another area where NMR spectroscopy is likely to advance our understanding of structure and function in hemoglobin is in determination of solution quaternary structure of the T state, using the residual dipolar coupling method. Although crystal structures of Hb in the deoxy form do not suggest as large a range of structures as in the distally ligated form,[15,16] studies of deoxy-Hb A trapped in sol–gel matrices suggest that there are at least two distinct highly populated conformations of hemoglobin in the T state that have significant affinity differences.[76] Residual dipolar coupling methods should be able to determine whether deoxy-Hb A in solution conforms well to the crystallographically observed T-quaternary structure, or whether additional structures are also present, as in distally ligated HbCO A.

Although hemoglobin has been studied with great intensity by a wide variety of methods over the last half-century, many of the molecular details that control allostery are still unknown. Experimental results on rHbs derived from NMR, equilibrium, and kinetic studies of ligand binding clearly indicate that hemoglobin is a rather flexible protein, and that its conformation can adapt to a variety of perturbations, for example, amino acid substitutions, binding of ligands, and allosteric effectors. These results also indicate that allosteric interactions in Hb A involve multiple pathways of signal transmission. It is only by uncovering the structural and conformational details of individual amino acid residues, their dynamics, and the structural heterogeneity of the Hb tetramer in solution that we can hope to understand the full details of hemoglobin allostery at atomic resolution.

Acknowledgments

Our hemoglobin research is supported by research grants from the National Institutes of Health (R01Hl-24525, P01HL-071064, and S10RR-11248 to C.H.), and by a grant-in-aid from the American Heart Association (AHA-SDG 9930126N to D.B.).

[76] N. Shibayama and S. Saigo, *FEBS Lett.* **492,** 50 (2001).

[3] Evaluating Cooperativity in Dimeric Hemoglobins

By Alberto Boffi and Emilia Chiancone

Introduction

A growing number of homo- and heterodimeric hemoglobins has been identified not only within several phyla of invertebrates and lower vertebrates, but also in unicellular microorganisms such as bacteria, algae, and fungi.[1–4] Even though all these proteins share the typical globin fold (with the exception of truncated hemoglobins in which two helices are missing), the quaternary assembly and the nature of the subunit interface differ widely. This observation, which implies that cooperative ligand binding arises from diverse molecular mechanisms, renders dimeric hemoglobins ideal molecular machineries for testing thermodynamic models and mechanistic hypotheses aimed at explaining the origin of cooperative phenomena.

The homodimeric hemoglobin from the mollusk *Scapharca inaequivalvis* (HbI) is by far the most extensively investigated protein within this class of proteins.[4] In HbI, cooperative ligand binding has been proposed to be entirely homotropic in origin as it occurs with minor quaternary structural rearrangements and no heterotropic effectors have been identified.[5] Thus, cooperativity is ascribed to the direct communication between the two heme groups that are in close contact across the subunit interface.[6] The heme propionate groups provide a direct pathway for heme–heme interaction[7] and ligand-linked structural changes are localized to the heme proximal side.

A completely different mechanism is observed in hemoglobins from lower vertebrates such as Cyclostomata, in which cooperative oxygen

[1] A. Fago, L. Giangiacomo, R. D'Avino, V. Carratore, M. Romano, A. Boffi, and E. Chiancone, *J. Biol. Chem.* **276,** 27415 (2001).

[2] M. E. Andersen, *J. Biol. Chem.* **246,** 4800 (1971).

[3] M. Couture, S. R. Yeh, B. A. Wittenberg, J. B. Wittenberg, Y. Ouellet, D. L. Rousseau, and M. Guertin, *Proc. Natl. Acad. Sci. USA* **96,** 11223 (1999).

[4] W. E. Royer, Jr., W. A. Hendrickson, and E. Chiancone, *Science* **249,** 518 (1990).

[5] M. Ikeda-Saito, T. Yonetani, E. Chiancone, F. Ascoli, D. Verzili, and E. Antonini, *J. Mol. Biol.* **170,** 1009 (1983).

[6] E. Chiancone, R. Elber, W. E. Royer, Jr., R. Regan, and Q. H. Gibson, *J. Biol. Chem.* **268,** 5711 (1993).

[7] D. L. Rousseau, S. Song, J. M. Friedman, A. Boffi, and E. Chiancone, *J. Biol. Chem.* **268,** 5719 (1993).

binding is the result of ligand-linked subunit dissociation. The behavior of lamprey *(Petromyzon marinus)* and hagfish *(Myxine glutinosa)* hemoglobins is prototypic: the ligand-bound species are monomeric, whereas the deoxygenated proteins are able to form both homodimers and heterodimers (and eventually higher order heteropolymers).[1,2,8] In these proteins, the subunit dissociation process that accompanies ligand binding is linked also to the release of Bohr protons. The coupling between these processes provides the entropic driving force for the observed cooperativity.[9]

This multifaceted behavior indicates there is no unique structural basis at the origin of cooperativity in dimeric hemoglobins, both from the thermodynamic and mechanistic points of view. Cooperative ligand binding thus appears to have been selected evolutionarily, despite the mutational pressure on the intersubunit contact regions (not conserved), as a consequence of a conserved tertiary fold endowed with sufficient conformational plasticity to allow ligand-linked, heme iron-driven structural rearrangements that ultimately lead to cooperative behaviors.

In this framework, it may be envisaged that cooperativity in dimeric hemoglobins can arise from two completely different mechanisms: (1) cooperative ligand binding driven by ligand-linked subunit dissociaton (mainly heterotropic), or (2) cooperative ligand binding within a stable dimer (homotropic). In real systems, the presence of one mechanism does not exclude the other.

Measurements of Monomer–Dimer Association–Dissociation Equilibria

Assessment of the existence of a finite monomer–dimer equilibrium among both liganded and unliganded proteins is a prerequisite for understanding the thermodynamic behavior and the kinetic properties of ligand binding in dimeric hemoglobins.

Methods for determination of the association state in solution comprise analytical ultracentrifugation, light scattering, and gel permeation techniques. Among these, analytical ultracentrifugation provides a direct, complete, and unequivocal description of the hydrodynamic properties of the protein in solution. It is the method of choice to establish the possible existence of a finite monomer–dimer equilibrium as it can be used both in the sedimentation velocity mode and in the sedimentation equilibrium mode. Further, the experiments are easy to perform and to analyze given the availability of computer-controlled analytical ultracentrifuges equipped

---

[8] M. E. Andersen and Q. H. Gibson, *J. Biol. Chem.* **246,** 4790 (1971).

[9] E. Chiancone and A. Boffi, *Biophys. Chem.* **86,** 173 (2000).

with absorbance optics (XL-A with a An60-Ti rotor Beckman Coulter, Fullerton, CA) and of dedicated software.

In analytical ultracentrifugation experiments, special care must be taken to avoid precipitation of unstable proteins or oxidation of the oxygenated derivative. Hemoglobin oxidation during the ultracentrifuge run might represent a major problem when investigating the association state of the oxygenated protein.[10] The heme iron oxidation-linked spectral changes may alter the equilibrium profile and the redox state-linked dimer dissociation may affect estimation of the association–dissociation constants. Carbon monoxide-bound adducts might be used instead of oxygenated derivatives to assess the association state of the ferrous, ligand-bound form. However, slight but sizeable molecular weight differences between the carbon monoxide and oxygenated derivatives are known to occur and have been reported, for example, for HbI.[11] When analyzing the deoxygenated derivative, it is essential to monitor accurately the possible reoxygenation of the sample due to oxygen leaking from the cell, especially when measuring equilibrium sedimentation of high oxygen affinity hemoglobins. The cells should be purged with nitrogen gas before loading the deoxygenated hemoglobin solution. Degassed sodium dithionite solutions (10–20 mM range, buffered at the desired pH) might be added to nitrogen gas-equilibrated protein solutions to ensure complete deoxygenation of the sample.

Sedimentation Velocity Measurements

The sedimentation velocity method allows fast and accurate determination of the association state of the protein provided that the association–dissociation equilibrium is rapidly reversible with respect to the rate of separation between monomer and dimer in the centrifugal field. The kinetic problem is common to all methods based on solute transport and can be completely overcome only by the use of thermodynamic equilibrium methods (see the next section). Given the molecular mass of the monomer (about 17 kDa), sedimentation velocity experiments are best carried out at 50,000–60,000 rpm. The widest possible range of protein concentration (1–500 μM) in the appropriate buffer should be explored. In the case of hemoglobins, the task is facilitated by the high molar absorptivity of the heme moiety ($>10^5$ M^{-1} cm^{-1}) that allows accurate measurements even at low

[10] C. Spagnuolo, F. Ascoli, E. Chiancone, P. Vecchini, and E. Antonini, *J. Mol. Biol.* **164,** 627 (1983).
[11] W. E. Royer, Jr., R. A. Fox, F. R. Smith, D. Zhu, and E. Braswell, *J. Biol. Chem.* **272,** 5689 (1997).

($<10^{-6}$ M) protein concentrations. Data are usually collected every 4 min at 410, 500, or 540 nm, depending on protein concentration, at a spacing of 0.005 cm with 3 averages in a continuous scan mode (at least 60 scans). The sedimentation profiles can be analyzed with the DCDT program (PC version provided by W. Stafford, Boston Biomedical Research, Boston, MA), which yields weight average sedimentation coefficients that can be corrected to $s_{20,w}$ using standard procedures.[12] The existence of a reversible monomer–dimer equilibrium manifests itself in the departure from the linear negative concentration dependence of the $s_{20,w}$ values, the norm for simple, noninteracting proteins. As discussed in detail in a previous volume of this series, this negative dependence, which is due to hydrodynamic factors, is counteracted by the effect of association such that, for relatively strong associations, the value of the sedimentation coefficient at first rises with an increase in protein concentration and then decreases.[13] The dependence of sedimentation coefficients on protein concentration can be fitted to a monomer–dimer equilibrium along the lines described in detail by Gilbert and Gilbert.[13]

Sedimentation Equilibrium Measurements

Sedimentation equilibrium experiments may yield accurate estimates of the equilibrium constants for any given hemoglobin derivative even in a single ultracentrifuge run (three samples at different protein concentrations). However, the long measurement times (1–2 days) required by equilibrium experiments may not always be compatible with the shelf life of the protein. In these instances sedimentation velocity should be preferred.

Sedimentation equilibrium runs are typically carried out at $10\,^{\circ}$C and between 30,000 and 40,000 rpm, and data are usually collected every 3–4 h at an appropriate wavelength (434 or 557 nm and 421 or 540 nm for deoxy- and oxyhemoglobin, respectively) at a spacing of 0.001 cm with 10 averages in a step scan mode. Equilibrium is checked by comparing successive scans. As a control, absorption spectra of all samples must be measured in the ultracentrifuge cell before and after the run. Data sets can be edited with REEDIT (J. Lary, National Analytical Ultracentrifugation Center, Storrs, CT) and fitted with NONLIN (PC version provided by E. Braswell, National Analytical Ultracentrifugation Center) according to Johnson et al.[14] Data from different concentrations and speeds can be combined for global fitting. For fits to a monomer–dimer association scheme, the monomer

---

[12] W. F. Stafford, *Anal. Biochem.* **203,** 295 (1992).
[13] L. M. Gilbert and G. A. Gilbert, *Methods Enzymol.* **27,** 273 (1973).
[14] M. Johnson, J. J. Correia, D. A. Yphantis, and H. Halvorson, *Biophys. J.* **36,** 575 (1981).

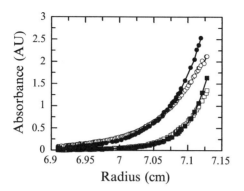

FIG. 1. Sedimentation equilibrium profiles of the HbI K30D mutant. Deoxygenated (●, ■) and CO-bound (○, □) derivatives analyzed in two separate runs at 30,000 (●, ○) and 40,000 (■, □) rpm. Data were obtained in 0.1 M phosphate buffer at pH 7.0 and 10 °C. The protein concentration was 280 μM. Continuous lines are fitted curves, which yield $K_{1,2}$ values of $(1.3 \pm 0.2) \times 10^3\ M^{-1}$ and $(8.0 \pm 0.1) \times 10^4\ M^{-1}$ for the CO-bound and deoxygenated derivative, respectively.

molecular weight can be fixed at the value determined from the amino acid sequence. The data shown in Fig. 1 provide an example of typical mono-mer–dimer equilibrium profiles as obtained for the liganded (CO bound) and unliganded derivatives of the HbI K30D mutant.[15]

Ligand-Binding Isotherms

Oxygen Equilibrium Measurements

Methods and techniques for oxygen equilibrium measurements in he-moglobins have been extensively reviewed in a previous volume of this series.[16] Therefore this chapter describes only those phenomenological working models that may allow discrimination between homotropic coop-erativity arising within stable dimers and cooperative ligand binding driven by protein association–dissociation phenomena. In the case of reversibly associating systems, the data should ideally cover a concentration range that permits description of the ligand-binding behavior of both monomers and dimers.

[15] P. Ceci, L. Giangiacomo, A. Boffi, and E. Chiancone, *J. Biol. Chem.* **277**, 6929 (2002).
[16] B. Giardina and G. Amiconi, *Methods Enzymol.* **76**, 417 (1981).

Data Analysis and Thermodynamic Models

Oxygen-binding curves of dimeric hemoglobins, reported as Hill plots, can be fitted according to different reaction schemes depending on whether the monomer–dimer equilibrium must be considered explicitly or not. Thermodynamic models that take into account reversible protein dimerization can be easily worked out, starting from the simple two-site Adair scheme of Fig. 2A.

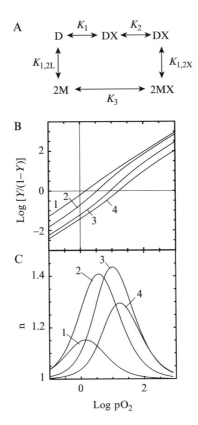

FIG. 2. Oxygen-binding isotherms as a function of protein concentration in a dimeric hemoglobin undergoing ligand-linked association–dissociation equilibrium. (A) Adair scheme for ligand-linked hemoglobin dimerization (see text for details). (B) Ligand-binding curves obtained according to Eqs. (1) and (2); (C) first derivative $\partial\{\log [Y/(1 - Y)]\}/\partial(\log \text{p}O_2)$ of each curve. Protein concentrations were as follows: (1) 1×10^{-3} M; (2) 1×10^{-4} M; (3) 1×10^{-5} M; (4) 1×10^{-6} M. Values of the thermodynamic parameters, which simulate the behavior of the *S. inaequivalvis* HbI K30D mutant, were as follows: $K_1 = K_2 = 0.03$ torr^{-1}; $K_3 = 1$ torr^{-1}; $K_{1,2D} = 1 \times 10^6$ M^{-1}; $K_{1,2L} = 1 \times 10^3$ M^{-1}.

In terms of the modified Adair model, the scheme reduces to the following equations:

$$Y = [K_3x + K_1K_2[M]xK_1K_2K_{1,2X}[M]x^2]/C \qquad (1)$$

$$C = [M] + K_3[M]x + 2K_1K_2[M]^2 2K_1K_2K_{1,2X}[M]^2x^2 \qquad (2)$$

where Y is the saturation fraction, $[M]$ is the concentration of the deoxygenated monomer, x is the equilibrium concentration of the ligand, and C is the total protein concentration. The thermodynamic constants K_1, K_2, and K_3 are the apparent oxygen-binding constants of the unliganded dimer, the monoligated dimer, and the unliganded monomer, respectively. $K_{1,2L}$ and $K_{1,2X}$ are the monomer–dimer association constants of the fully oxygenated and unliganded protein, respectively.

The data sets, obtained as a function of protein concentration, can be arranged into a single matrix and fitting curves relative to each binding isotherm can be generated at each iteration. Least squares are thus calculated on the whole data set (global fitting procedure) by minimizing the values of the fitting parameters to Eq. (1), namely, the oxygen-binding constant of the unligated species (K_1), the oxygen-binding constant of the monoliganded species (K_2), and the monomer–dimer equilibrium constants of the unliganded and diliganded derivatives ($K_{1,2L}$ and $K_{1,2X}$). In our experience, the values of $K_{1,2L}$ and $K_{1,2X}$ obtained from ultracentrifugation data, used as initial guesses for the global fitting procedure, may not permit unambiguous determination of the other parameters. The confidence limit for the fitted values of K_1 and K_2 can be increased and a possible bias singled out by fixing the value of $K_{1,2L}$ or $K_{1,2X}$ in independent curve-fitting procedures.

The simulation presented in Fig. 2, which mimics the behavior of the HbI K30D mutant,[15] brings out the essential features of the modified Adair scheme as evident from the Hill plot (Fig. 2B) and its first derivative (Fig. 2C). The binding isotherms as a function of protein concentration (from 1 μM to 1 mM) demonstrate the occurrence of cooperative ligand binding even in the absence of homotropic interactions ($K_1 = K_2$) over a protein concentration range in which dimers are present in the unliganded species ($K_{1,2L} = 10^7 \, M^{-1}$) and are in negligible amount in the liganded form ($K_{1,2X} = 10^3 \, M^{-1}$). Under these conditions, cooperativity appears as an entirely entropy-driven phenomenon, dominated by the gain of translational entropy occurring on deoxygenation. On the other hand, under conditions in which the protein is a stable dimer in both the liganded and unliganded derivatives, the presence of cooperativity is dominated by a homotropic interaction with $K_2 > K_1$.

It should be pointed out that the modified Adair scheme, although endowed with high flexibility and a good predictive value, rests on a purely phenomenological basis. It serves essentially to guide the thermodynamic analysis toward discrimination between intradimer cooperativity (homotropic) and subunit association-linked cooperativity.

More symmetric models may be used in order to single out the free energy involved in the conformational changes at the basis of cooperative ligand binding. The classic two-state Monod–Wyman–Changeux (MWC) model, however, cannot describe adequately homotropic cooperativity within stable dimers. In fact, the three parameters L, K_R, and K_T are redundant for a system whose behavior can be accounted for by two parameters only. The cooperon model,[17] which integrates features of both the induced fit and of the two-state model, has been applied successfully to describe the cooperative oxygen binding in the dimeric hemoglobin from the mollusk *Nassa mutabilis*.[18]

Although the use of pure thermodynamic models may provide valuable insight into the energetics of the cooperative behavior, the present goal, imposed by the massive release of structural data obtained by X-ray crystallography and nuclear magnetic resonance (NMR) spectroscopy, is the understanding of cooperativity in terms of statistical thermodynamic models that may take into account explicitly single bond formation/cleavage. The case of HbI is again an illuminating one. The crystal structures of unliganded and ligand bound HbI reveal that ligand binding is accompanied by the release of a set of highly ordered water molecules from the subunit interface to the bulk solvent.[9,11] Ligand-linked release of water is accompanied by an entropy increase that may provide a sizeable contribution to the overall cooperative free energy. Thus, homotropic cooperativity in HbI could have a heterotropic component if ligand-linked water molecules were considered as allosteric effector molecules.

Kinetic Manifestations of Cooperativity

Kinetics of Oxygen Release

Analysis of the time courses of oxygen dissociation provides key information about the mechanism of cooperativity in dimeric hemoglobins. In fact, the kinetic counterpart of cooperativity is manifest in the oxygen release process rather than in the combination process. The reverse is observed in the case of carbon monoxide, where cooperativity is apparent in the binding process. No exception to this general rule has ever been reported for any hemoglobin.

[17] M. Brunori, M. Coletta, and E. Di Cera, *Biophys. Chem.* **23**, 215 (1986).
[18] M. Coletta, P. Ascenzi, F. Polizio, G. Smulevich, R. del Gaudio, M. Piscopo, and G. Geraci, *Biochemistry* **37**, 2873 (1998).

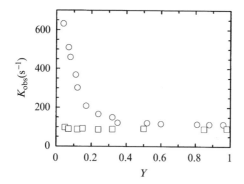

FIG. 3. First-order rate constants of oxygen release as a function of oxygen saturation in native *S. inaequivalvis* HbI and in the K30D mutant. Data were obtained in 0.25 *M* phosphate buffer at pH 7.0 and 20 °C by mixing the oxygenated protein at a concentration of 10 μM heme in the presence of 1–100 m*M* sodium dithionite with oxygen-containing buffer. The time courses, obtained by monitoring the absorbance decrease at 434 nm were fitted to monoexponential curves. The native HbI (○) and K30D mutant (□).

Oxygen release kinetics are best monitored in oxygen pulse experiments carried out in a stopped-flow apparatus by mixing the protein solutions (2–200 μM) in the presence of excess sodium dithionite (0.02–0.1 *M*) with buffer solutions containing oxygen at concentrations varying from 1400 to 2 μM. Oxygen binds to the heme iron within the dead time of the instrument (1–5 ms) and is released subsequently due to the rapid reaction of oxygen itself with dithionite to produce higher sulfur oxides.[19] Thus, the observed spectral signal pertains to the formation of deoxy heme and is monitored at the relevant Soret peak around 434 nm. Time courses are dithionite concentration dependent up to concentrations at which the rate of oxygen release becomes the rate-limiting process. It is thus essential, before setting up the oxygen pulse experiment, to determine carefully the dithionite concentration values at which the oxygen release process becomes dithionite concentration independent. The measured first-order rate constants are plotted as a function of oxygen saturation obtained by normalizing the observed amplitude at 434 or 557 nm, to the measured oxy-minus-deoxy spectrum (see Fig. 3). Although the oxygen pulse experiment is a simple one, researchers must be warned of several possible artifacts that may lead to a serious misinterpretation of the experimental data. In fact, binding of dithionite by-products (possibly tetrathionate ions or hydrogen peroxide) to the heme iron is not uncommon[19] and may lead to the

[19] J. S. Olson, *Methods Enzymol.* **76,** 631 (1981).

formation of ferrous heme iron adducts (hemochromogen). Thus, the stability of the deoxy hemoglobin spectrum in the presence of excess dithionite must be checked carefully before and after the pulse experiments.

Typical results of an oxygen pulse experiment are depicted in Fig. 3 and show, at least qualitatively, the different behavior of stable cooperative dimers (native HbI) versus a dimeric hemoglobin in which cooperativity results from a ligand-linked subunit association process (HbI K30D mutant). In native HbI, the observed first-order rate increases from 100 s^{-1} at high oxygen saturation, at which the measurement monitors oxygen release from the liganded conformation (R-like state), to 600 s^{-1} at low oxygen saturation, at which the measurement reflects primarily release from the monoligated species (T-like state). Thus, the increase in the rate of oxygen release at low oxygen saturation can be considered the kinetic manifestation of heme–heme interaction.[15] In contrast, no increase in the rate of oxygen release is observed in the HbI K30D mutant even at low oxygen saturation. For this mutant, protein cooperativity arises from oxygenation-induced monomerization and homotropic cooperativity is essentially abolished. This finding in turn implies that the kinetics of protein dissociation is so fast that it does not affect or limit the rate of oxygen release. More complex kinetic behaviors can be expected when performing oxygen pulse experiments on dimeric hemoglobins undergoing ligand-linked association–dissociation processes if the rate of oxygen release and that of protein dissociation are of comparable magnitude.

[4] Measuring Assembly and Binding in Human Embryonic Hemoglobins

By Thomas Brittain

Introduction

Although adult human hemoglobin is the most thoroughly studied protein and arguably one of the best understood, in terms of its function, the methods and models developed to investigate the detailed actions of this protein have not been widely applied to other hemoglobin systems. This chapter seeks to explore the methods and results obtained from the application of a range of techniques used to measure the assembly and binding processes that occur in the human embryonic hemoglobins. The human embryonic hemoglobins consist of a group of three hemoglobins synthesized in yolk sac-derived megaloblasts from approximately week 3 to week

16 of gestation. The three hemoglobins are tetrameric in nature and contain at least one uniquely embryonic globin chain: Gower I $\zeta_2\varepsilon_2$, Gower II $\alpha_2\varepsilon_2$, and Portland $\zeta_2\gamma_2$. In the wider context, this chapter serves to illustrate the fact that, in order to assign the biological activity of a novel hemoglobin system, it is necessary to apply a wide range of methodologies. Only then can a proper understanding of this complex protein, which shows both ligand and redox sensitivity within a multisubunit allosteric context, be obtained.

Specifically, the methodologies described in this chapter have been applied successfully to the study of the three human embryonic hemoglobins obtained from recombinant yeast expression systems. Each of these hemoglobin expression systems consists of *Saccharomyces cerevisiae* cells containing plasmids based on the naturally occurring yeast 2μ plasmid. The plasmids have been engineered to contain each of the appropriate globin genes, linked in tandem to artificial galactose promoters. Nevertheless, the approaches described here are, in principle, applicable to any hemoglobin system.

Heme–Globin Interactions

Heme–Globin Assembly In Vivo

In adult humans, hemoglobin assembly occurs in the reticulocyte when the gene products of the α- and β-globin genes combine with heme to form the $\alpha_2\beta_2$ tetrameric form of the protein. An analogous process is presumed to occur in recombinant expression systems. To obtain detailed information relating to the assembly process it is necessary to have a means to detect not only the product of the reaction but also the presence of intermediates. Nuclear magnetic resonance (NMR) spectroscopy provides the necessary detection system. It has long been recognized that, *in vitro,* rapidly mixing apoglobin proteins with an excess of heme leads to the formation of a hemoglobin tetramer within milliseconds.[1] Although the initial kinetic product of this reaction shows all the major reactivities of normal adult hemoglobin it also exhibits some minor structural differences. In particular, the heme group shows rotational heterogeneity, produced by 180° rotation about the heme α,γ-meso axis, in the initial kinetic product, which decays to a single predominant thermodynamic product identical with that present in mature red blood cells.[2] The sequential reaction of semiglobin

[1] M. Y. Rose and J. S. Olson, *J. Biol. Chem.* **258,** 4298 (1983).
[2] G. N. La Mar, Y. Yamamoto, T. Jue, K. M. Smith, and R. K. Pandey, *Biochemistry* **24,** 3826 (1985).

with limiting amounts of heme likewise leads to a discernibly different pattern of heme rotamer heterogeneity in the initial kinetic product, which slowly decays to the natural distribution in the thermodynamic product.[3] This rotational heterogeneity can be identified by NMR spectroscopy and can be used to monitor the assembly process in both natural and recombinant systems. An investigation of hemoglobin assembly by this approach involves a number of steps. First, hemoglobin-synthesizing cells must be sampled at appropriate time intervals when the hemoglobin must be trapped in a nonrelaxing form. The hemoglobin is then isolated and converted into a form appropriate for NMR study. NMR data are then collected and the rotamer distributions compared with those obtained *in vitro* under known conditions.

Hemoglobin Sampling. For recombinant systems, cells synthesizing hemoglobin may be trivially sampled at various times postinduction of the globin genes. In the case of red blood cells it is necessary to obtain circulating reticulocytes. In either case it is important to prevent decay of the heme heterogeneity over the period involved in hemoglobin isolation (see below). Reticulocytes may be obtained from freshly drawn peripheral blood samples by the method of differential density centrifugation.[4]

Trapping. To prevent decay of the heme rotational heterogeneity, over the time period of experimentation, it is necessary to trap the hemoglobin in a stable form.[2] For our purposes we find that trapping in the ferrous carbonmonoxy form is appropriate. This can be achieved simply by bubbling the cell suspension with carbon monoxide immediately after the cells are isolated.

Hemoglobin Isolation. Once trapped in the carbonmonoxy ferrous form, hemoglobin isolation can proceed as for the oxy form of the protein, using standard protocols, with the single modification that all buffers used are preequilibrated by bubbling with carbon monoxide.

Preparation of Hemoglobin Samples for NMR Spectroscopy. When preparing samples of hemoglobin for study by NMR it is important to recognize two important facts relating to the nature of the sample used. First, NMR is intrinsically an insensitive form of spectroscopy. This means that the hemoglobin sample needs to be at a concentration of 0.5 to 1.0 mM in order to obtain reasonably noise-free spectra (a volume of approximately 0.5 ml is sufficient). Second, NMR spectroscopy produces spectra that are extremely information rich—that is, they contain many signals. For the purposes of these experiments it is useful to ensure that the signals indicative of heme rotational heterogeneity fall outside the envelope due to most of

[3] K. Ishimori and I. Morishima, *Biochemistry* **27**, 4747 (1988).
[4] M. P. Sorette, K. Shiffer, and M. R. Clark, *Blood* **1**, 249 (1992).

the other proton signals generated by the protein. This can be achieved by ensuring the hemoglobin is in the appropriate ligand and redox state. The protein should be in the ferric state and we find the azide ligand suitable for these studies. (The interested reader may refer to Refs. 5 and 6 for a deeper understanding of the theoretical basis of the ligand and redox requirements for these experiments.) The NMR sample can be created directly from the carbon monoxide-trapped form by photolysis of the sample, in the presence of a small molar excess of potassium ferricyanide and an excess of sodium azide, using a standard 50-W bulb. (The exact excess of azide required should be determined by prior optical titration of the hemoglobin under study[7]; typically a 20- to 40-fold molar excess has been found sufficient in the case of the human adult and embryonic hemoglobins.) The sample is prepared in 25 mM Tris buffer containing 25 mM NaCl at pH 7.5 with 10% deuterium oxide and 2,2-dimethyl-2-sila-pentane-5-sulfonate, to facilitate chemical shift referencing.

NMR Data Collection. Using protein samples such as those described above, in a typical 400-MHz NMR spectrometer, it is possible to collect useful spectra in a relatively short time. Typically spectra can be obtained with 2000–50,000 transients (depending on protein concentration) with a spectral width of 20 MHz, 36,000 data points, a 6.5-μs 90° pulse, and a cycle time of 0.4 s. It should be noted that it may be necessary to perform the spectroscopic investigation over a range of temperatures in order to identify the temperature at which the spectra are best resolved. In general, higher temperature gives better resolution. However, the peak positions are themselves temperature dependent and it is not possible per se to predict the optimum temperature. In our studies 28–30° has been found to be optimum.

Spectral Analysis. The NMR signals located in the region of −12 to −30 ppm arise from the heme side chains and are indicative of rotational heterogeneity (Fig. 1). The two possible rotamers lead to two sets of NMR signals. For the α and β chains the two set of signals arising from the two possible rotamers have been assigned.[8] A serial comparison of the spectra of the embryonic hemoglobin tetramers with the spectrum of adult protein has allowed the assignment of the rotamer signals for each of the embryonic hemoglobin chains. Should this method be used for other hemoglobin systems it would be necessary first to assign each of the proton resonances to a particular chain and then to identify which of the rotamer

[5] I. Bertini, C. Luchinat, G. Parigi, and F. A. Walker, *J. Biol. Inorg. Chem.* **4,** 515 (1999).

[6] N. V. Shokhirev and F. A. Walker, *J. Biol. Inorg. Chem.* **3,** 581 (1998).

[7] T. Brittain, *J. Inorg. Biochem.* **81,** 99 (2000).

[8] G. N. La Mar, H. Toi, and R. Krishnamoorthi, *J. Am. Chem. Soc.* **106,** 6395 (1984).

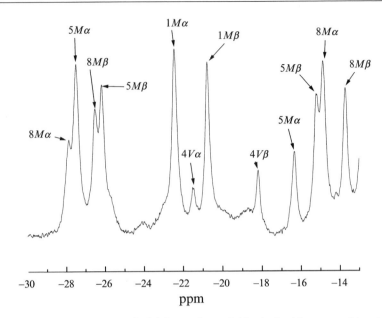

FIG. 1. The NMR spectrum of adult human hemoglobin obtained from recombinant yeast cells. The proton resonances labeled M (methyl) and V (vinyl) originate from different heme side chains. The resonances indicated in boldface type correspond to those from the heme rotamer not seen in the crystal structure of the thermodynamically stable form of the protein. Experimental conditions were those given in text.

signals were present in the thermodynamic product of heme binding. The level of disorder is simply defined by the ratios of these signal areas to the total signal area, arising from a particular side chain. It is therefore necessary to obtain the peak areas for the various signals and this task can be performed by many modern software packages capable of determining peak areas from incompletely resolved spectra. We have found Peakfit (SPSS, Chicago, IL) particularly useful for this purpose.

It has previously been shown that combination of globin protein with either stoichiometric or excess heme immediately leads to the production of hemoglobin containing essentially equal amounts of the two possible heme rotamers.[8] Reaction of globin with a substoichiometric amount of heme gives heme insertion into the α chain.[9,10] Further reaction of this hemilobin gives a hemoglobin molecule containing $\sim 0\%$ heme disorder in the β chain.[3] In contrast, circulating adult hemoglobin shows 2% heterogeneity in the α chain and 10% in the β chain. A different pattern of heterogeneity found in

[9] K. H. Winterhalter and D. A. Deranleau, *Biochemistry* **6**, 3136 (1967).
[10] G. Vasudevan and M. J. McDonald, *J. Biol. Chem.* **272**, 517 (1997).

both the newly synthesized adult protein obtained from reticulocytes and in recombinantly expressed embryonic hemoglobin has led to the proposal that *in vivo* hemoglobin synthesis occurs cotranslationaly,[11] a view that is supported by earlier radioactive labeling experiments.[12]

Interestingly, the human embryonic hemoglobins synthesized in yeast cells exhibit, in the initial kinetic product, approximately 25% disorder in their β chains, a value that equals that present in the thermodynamic distribution ratio. Adult human hemoglobin newly synthesized in yeast cells exhibits 32% α and 45% β heme disorder.[11]

Heme–Globin Affinity

Having identified the mechanism of heme insertion into globin proteins *in vivo* it is also possible to probe the affinity with which heme is bound, using *in vitro* measurements. Note, however, that heme binding to globins is conveniently measured only with the ferric form of heme, as the liganded ferrous forms have an extremely high binding affinity. A convenient method to determine the affinity of hemin for a particular globin is to monitor the partitioning of the hemin between the native globin and a suitable alternative acceptor. For this procedure to be viable two major requirements need to be met by the acceptor, namely, (1) the acceptor must exhibit a hemin-binding affinity of similar magnitude to the native globin, such that neither excessive concentrations of acceptor are required to produce a reasonable level of transfer nor simple stiochiometric transfer occurs and (2) the transfer of hemin must be associated with a measurable change in some observable property of the system. Benesch and Kwong[13,14] have demonstrated that serum albumin fulfills both these requirements. The hemin affinity of human serum albumin (approximately picomolar) makes its use appropriate for the study of heme binding to human adult hemoglobin and other globins of somewhat higher heme affinity, whereas bovine serum albumin has a heme-binding affinity (approximately nanomolar) suitable for use in measuring heme binding to globins of lower affinity. In both cases heme transfer at alkaline pH can be easily monitored by observing the changes in visible absorption spectra, over a wide range of protein concentrations. The application of these partitioning methods is not only simple but also gives access to information relating to hemoglobin

[11] A. J. Mathews and T. Brittain, *Biochem. J.* **357,** 305 (2001).
[12] A. A. Komar, A. Kommer, I. A. Krasheninnikov, and A. S. Spirin, *FEBS Lett.* **326,** 261 (1993).
[13] R. E. Benesch and S. Kwong, *J. Biol. Chem.* **265,** 14881 (1990).
[14] R. E. Benesch and S. Kwong, *J. Biol. Chem.* **270,** 13785 (1995).

dimerization (see below). The methods are, however, generally limited to studies of heme binding to the "β-type" chains as the "α-type" chains of most hemoglobins exhibit a much tighter binding of heme. In the case of heme binding to human embryonic hemoglobins the later restriction does not pose a significant problem, however, as we find that high levels of human serum albumin, which partially depletes α-type chains of heme, leads to precipitation of apoglobin products, which in itself would preclude measurement of heme binding to the α-type chains.

Experimental

As the spectral properties of ferric hemoglobin and metheme–albumin are temperature, pH, and ionic strength dependent it is necessary to calibrate the spectral properties of both substances at a chosen pH, temperature, and ionic strength. Then, using a fixed hemoglobin concentration, absorption measurements are made at various concentrations of human serum albumin. For studies of heme binding to the human embryonic hemoglobins we typically measure spectra of 5 μM hemoglobin in the presence of albumin ranging in concentration from 0.1 to 25 molar equivalents, in 100 mM Tris buffer, pH 9.0 at 20°. The hemoglobin sample is prepared from the carbonmonoxy form of the protein by photolysis in the presence of a 2-fold molar excess of potassium ferricyanide, followed by the removal of low molecular weight products by passage down a small Sephadex column equilibrated with the appropriate buffer. After preparation of the mixture, equilibration typically requires 10 min under these conditions. Note, however, that the spectral characteristics of the hemoglobin/metheme–albumin system are such that measurements can easily be made using hemoglobin concentrations from 1 μM (measuring in the 400-nm region of the spectrum) to 100 μM or more (using measurements in the 500- to 600-nm region).

On the basis of the calibration spectra, results have often been reported in terms of the equilibrium distribution ratio *(R)* given by

$$R = \frac{[\text{metheme–albumin}][\text{apohemoglobin}]}{[\text{methemoglobin}][\text{albumin}]} \tag{1}$$

However, knowing the absolute value of the albumin–heme binding constant (bovine serum albumin, 4 × 10^9 M^{-1}; human serum albumin, 4 × 10^{12} M^{-1}) the binding constant for hemoglobin can be obtained by a simple ratio.

When applied to the human embryonic hemoglobins we find that the embryonic ε and γ chains bind heme 25 and 10 times more tightly, respectively, than do the β chains.[15] Structural studies have indicated that this increase in heme affinity probably arises from the unusual conformation of Lys-66 in the ε and γ chains, which produces a stabilizing interaction with the heme propionic acid side chains.

Although the binding constant for heme could also, in principle, be obtained from the ratio of the independently determined association and dissociation rate constants, the complexity of the binding process for heme to tetrameric hemoglobin normally makes this impractical.

Subunit Interactions: Protein Dissociation

In solution equimolar quantities of α- and β-globin proteins self-assemble to yield fully functional $\alpha_2\beta_2$ tetrameric proteins on the millisecond time scale. Under conditions that might pertain physiologically only the $\alpha_2\beta_2 \leftrightarrow 2\alpha\beta$ equilibrium is of significance. Furthermore many accumulated data indicate that the equilibrium constant for this dissociation process, in the adult human hemoglobin, is sensitive to the oxygenation state of the globin subunits, being typically in the micromolar region for the oxy form and in the low nanomolar region for the deoxy form of the protein. In principle, the dimerization constant ($K_{4,2}$) can be determined by observation of any property, as a function of hemoglobin concentration, that is different for the tetrameric and dimeric forms of the protein. Clearly the most precise data will be obtained from experiments in which the properties of the tetramer and dimer are most different.

Oxy Form

As most liganded hemoglobins dimerize in the micromolar range, direct measurements of the position of the equilibrium are possible using ultraviolet–visible (UV–Vis) spectroscopy. Two of the most sensitive measures are heme transfer and ligand-binding rates. Although heme transfer is measured on the ferric form of the protein, many data have been accumulated that the ferric protein acts analogously to the liganded ferrous protein with regard to dissociation. In practice it has been shown that the rate of heme loss from the β-type subunits is sensitive to the state of aggregation of the subunits. Furthermore, as the rates of the association and dissociation processes are rapid compared with the rate of heme transfer, the observed rate of heme transfer is simply the weighted sum of the rates due to dimer and tetramer reactivity according to

$$k_{obs} = k_dD + k_t(1 - D) \qquad (2)$$

[15] N. Robson and T. Brittain, *J. Inorg. Biochem.* **64**, 137 (1996).

where k_{obs} is the observed rate of heme transfer, k_d and k_t are the limiting rates for dimer and tetramer reactivity observed, respectively, at very low and very high protein concentrations, and D is the fraction of hemes present in the dimeric form of the protein. D is given by

$$D = \left\{-K_{4,2} + \left\{(K_{4,2})^2 + 4[\text{Hb}]K_{4,2}\right\}^{1/2}\right\}/2[\text{Hb}] \qquad (3)$$

The values of $K_{4,2}$, k_d, and k_t can then be determined by nonlinear least-squares fitting of the observed rate of heme transfer as a function of total hemoglobin concentration. Appropriate experimental data can be obtained using the basic protocol described above for the measurement of heme affinity. However, in these measurements, rather than using a fixed hemoglobin concentration and varying the ratio of human serum albumin present, a constant ratio of human serum albumin (2:1) is employed and the concentration of hemoglobin is varied. Furthermore, rather than determine the equilibrium position, the initial rate of reaction is determined, subsequent to the mixing of the albumin and hemoglobin solutions. The advantage of using albumin for the transfer reaction, rather than the more avid green Tyr mutant of myoglobin,[16] arises from the fact that removal of the heme only from the β-type chains does not lead to significant precipitation of apoglobin, which would otherwise occur. Also, only the rate of β-chain heme removal is aggregation dependent and so α-chain data would be redundant. By employing standard pathlength cells and combining measurements in the 400-nm region (at low hemoglobin concentrations, <5 μM) and in the 500- to 600-nm region (at high protein concentrations, >5 μM), it is possible to make direct determinations over three magnitudes of hemoglobin concentration. Such data are suitable for investigating hemoglobins with $K_{4,2}$ values in the region of 0.1 to 20 μM. For systems with lower dissociation constants the use of longer pathlength cells in the 400-nm region allows measurements down to $\sim$0.01 μM.

A major limitation on the use of heme transfer to albumin is that measurements are only viable at alkaline pH, although at least for adult human hemoglobin the evidence is that $K_{4,2}$ varies little from pH 9.0 to pH 7.4. However, we cannot assume the same degree of pH insensitivity in all hemoglobins and so we have employed another high-sensitivity technique to test this in the human embryonic hemoglobins, although this method requires access to much more sophisticated apparatus. The dimerization constant for hemoglobin can also be determined by measurement of ligand recombination after photolysis. For the purpose of these experiments the

[16] M. S. Hargrove, E. W. Singleton, M. L. Quillin, L. A. Ortiz, G. N. Phillips, J. S. Olson, and A. J. Mathews, *J. Biol. Chem.* **269**, 4207 (1994).

ligand of choice is carbon monoxide (although interest is normally in the functioning of the oxygenated form of the protein, carbon monoxide has been shown to exhibit the same reactivity as oxygen, modified by a scaling factor, and has more useful photochemistry). At neutral pH recombination of photolysed hemoglobin with carbon monoxide occurs as the sum of two reactions. Fast recombination occurs with the dimeric form of the protein and slow recombination occurs with the tetrameric form. Thus the measurement of the fractional contribution made by the faster reaction gives a direct indication of the fraction of dimer present. A nonlinear least-squares fit of the fraction of dimer as a function of hemoglobin concentration according to Eq. (3) then gives the value of $K_{4,2}$.

In typical experiments we employ 100 mM Bis-Tris buffer at pH 6.5–7.5 equilibrated with 100 μM CO at 20°. Using a conventional microsecond flash lamp photolysis system (Applied Photophysics, Leatherhead, UK) with various pathlength cells and monitoring the reaction at 435 nm, the recombination reaction occurs on a time scale of tens of milliseconds and it is feasible to investigate the recombination over a range of hemoglobin concentrations (~0.1–100 μM).

Application of the heme exchange rate technique to the three human embryonic hemoglobins shows that in all cases the observed rate of β-type chain exchange is sensitive to the concentration of hemoglobin. Analysis of the data according to Eqs. (2) and (3) indicates that the dimerization of the oxygenated forms of the embryonic hemoglobins is associated with equilibrium constants similar to that of the adult protein at pH 9.0 (Table I).

Ligand recombination studies applied to measurements of the dimerization of the liganded forms of the human embryonic hemoglobins at pH 6.5 yield values of $K_{4,2}$ that are similar to those obtained at pH 9.0, using the heme transfer technique[17] (see Table I). These results indicate that in common with adult human hemoglobin the embryonic hemoglobins show little sensitivity in the dimerization reaction of the oxygenated form of the protein between pH 6.5 and 9.0. (Note, however, that at low pH (pH 4.9), at least hemoglobin Portland ($\zeta_2\gamma_2$) undergoes significant dissociation and reassociation of subunits to yield the more stable hemoglobin Barts (γ_4).[18]

Deoxy Form

The deoxy form of hemoglobin is usually significantly more resistant to dissociation than is the oxy form, with a $K_{4,2}$ for the adult protein of

[17] O. Hofmann and T. Brittain, *Biochem. J.* **315,** 65 (1996).

[18] R. Kidd, A. J. Mathews, H. M. Baker, T. Brittain, and E. N. Baker, *Acta Crystallogr. D Biol. Crystallogr.* **57,** 921 (2001).

TABLE I
TETRAMER–DIMER EQUILIBRIUM CONSTANTS FOR OXYHEMOGLOBINS

Protein	$K_{4,2}$	
	Heme exchange[a]	Ligand recombination[b]
$\alpha_2\beta_2$	1.5	2.0
$\alpha_2\varepsilon_2$	1.0	2.5
$\zeta_2\varepsilon_2$	1.5	2.0
$\zeta_2\gamma_2$	4.0	3.4

[a] $K_{4,2}$ (micromolar) from heme exchange experiments[15] at pH 9.0.
[b] $K_{4,2}$ (micromolar) from carbon monoxide photolysis experiments[17] at pH 6.5.

2×10^{-8} to $4 \times 10^{-9} M$ at pH 9.0.[19,20] Such low values of $K_{4,2}$ normally pre-clude the direct observation of equilibrium properties in this system. How-ever, the dimerization constant can be estimated from the ratio of the dissociation and association rate constants. To measure the appropriate rate constants we make use of the fact that, at micromolar concentrations, a solution of liganded hemoglobin contains an appreciable concentration of dimers, whereas at the same concentration, a solution of deoxy hemoglobin is essentially devoid of dimers. Thus the association rate of dimers can be determined by monitoring the association process following the rapid de-oxygenation of a micromolar solution of oxyhemoglobin. This reaction can be achieved in a stopped-flow apparatus by rapidly mixing 1–5 μM oxy-hemoglobin with a solution containing an excess of sodium dithionite and monitoring the reaction time course at 430 nm. Some caution, however, is necessary concerning the exact conditions employed. Sodium dithionite is air sensitive and its oxidation products can lead to the production of arti-facts in the reaction. The sodium dithionite solution should thus be kept to a reasonably low concentration, still commensurate with rapid deoxy-genation of the hemoglobin sample. It is also necessary to recognize that sodium dithionite is a reasonably strong acid. In practice we add solid sodium dithionite anaerobically to a thoroughly degassed and nitrogen-equilibrated buffer of 100 mM Bis-Tris at pH 7.4 to give a final concentra-tion of 2 mg/ml. To reduce oxidation artifacts we also routinely equilibrate the hemoglobin solution with oxygen to a concentration commensurate with full oxygenation but not oxygen saturation. For high oxygen affinity hemoglobins such as the human embryonic hemoglobins this may be as

[19] D. H. Atha and A. Riggs, *J. Biol. Chem.* **251**, 5537 (1976).
[20] A. H. Chu and G. K. Ackers, *J. Biol. Chem.* **256**, 1199 (1981).

low as 50 μM oxygen. Under these conditions the reaction traces obtained at 430 nm consist of two processes. The initial rapid deoxygenation reaction occurs in less than 1 s whereas the dimer association reaction occurs on a time scale of tens of seconds. The dimer association is purely second order and so, to determine the association rate constant, it is necessary to have an accurate measure of the concentration of dimers initially present. The initial deoxy dimer concentration is taken to be half that present in the oxygenated sample, before mixing, and is calculated from the known hemoglobin concentration and the measured $K_{4,2}$ value (see above). A range of hemoglobin concentrations should be used to confirm the order of the reaction observed.

The rate of deoxy tetramer dissociation rate can be determined by monitoring the rate-limited reaction of deoxy dimers with haptoglobin. Haptoglobin binds dimeric forms of hemoglobin with high affinity ($\sim 10^{15} M$). Mixing deoxyhemoglobin, at micromolar concentrations with a slight excess of haptoglobin will therefore lead to two reactions. The trace equilibrium amount of deoxy dimers present will immediately bind to haptoglobin. Subsequent slow production of deoxy dimers arising from dissociation of deoxy tetramers will then rate limit further reaction with haptoglobin. The haptoglobin–dimer complex has a slightly lower extinction ($\sim 17\%$) than the free dimers at 430 nm and therefore observation at this wavelength can be used to monitor, indirectly, the dissociation process.

Typically the reaction can be carried out in stoppered spectrophotometer cuvettes. The reaction is initiated by the anaerobic injection of the appropriate, small volume of a concentrated deoxygenated hemoglobin sample into the anaerobic cuvette, containing haptoglobin in the presence of a small excess of sodium dithionite, to give approximately 10 μM hemoglobin and 20 μM haptoglobin. When the reaction is performed at pH 9.0 in 100 mM Tris buffer the reaction is essentially complete in 20 min. min. Although at neutral pH the reaction of haptoglobin with hemoglobin has variously been reported as both monophasic and biphasic,[20,21] in our experience at alkaline pH the reaction is purely first order.

When the deoxygenated forms of the three human embryonic hemoglobins are investigated by these two methods each protein shows a common pattern of reactivity. The rates of dissociation of the deoxygenated tetrameric forms of the proteins are similar in all cases, with rate constants of 1–2 $\times$ 10^{-3} s^{-1}. Although the pattern of deoxy dimer association is common to all the proteins the association rate constants shows significant differences. The ε chain-containing embryonic hemoglobins exhibit dimer association rate constants similar to that observed for the adult protein

[21] S. H. C. Ip, M. L. Johnson, and G. K. Ackers, *Biochemistry* **15,** 654 (1976).

TABLE II
TETRAMER–DIMER EQUILIBRIA FOR DEOXYHEMOGLOBINS

Protein	Association rate[a]	Dissociation rate[b]	$K_{4,2}{}^{c}$
$\alpha_2\beta_2$	19.5	1.3	1.5
$\alpha_2\varepsilon_2$	18.4	1.5	1.2
$\zeta_2\varepsilon_2$	14.5	1.9	0.75
$\zeta_2\gamma_2$	2.2	1.1	0.2

[a] Rate determined from stopped-flow deoxygenation of oxygenated hemoglobin ($\times 10^{-4}\ M^{-1}\ s^{-1}$).
[b] Rate determined from reaction with haptoglobin ($\times 10^{3}\ s^{-1}$).
[c] Value obtained from the ratio of association–dissociation rate constants ($\times 10^{-8}\ M^{-1}$).

under identical conditions, with an association rate constant of $1.5–2.0 \times 10^{5}\ M^{-1}\ s^{-1}$. The human embryonic hemoglobin Portland ($\zeta_2\gamma_2$), however, exhibits a much smaller association rate constant of $2.2 \times 10^{4}\ M^{-1}\ s^{-1}$ and hence a significantly smaller equilibrium constant of $2 \times 10^{7}\ M^{-1}$ (Table II). Direct evidence of this different association constant is seen in the observed tendency for hemoglobin Portland to undergo subunit exchange at low pH.[18]

Hemoglobin–Protein Interactions

Although the normal physiological functioning of hemoglobin involves mainly the interaction of the hemoglobin protein tetramer with various small molecules, the maintenance of the physiological role of hemoglobin requires regular interactions with other proteins within the red blood cell, in particular cytochrome b_5. This interaction is crucial to the maintenance of the heme iron in the oxygen-binding ferrous state as, during normal circulation, approximately 2% of the total hemoglobin, within an adult red blood cell, is oxidized to the nonphysiologically active ferric form each day. Electron transfer from cytochrome b_5 to hemoglobin serves to reverse this process and return the hemoglobin to the physiologically active ferrous form. These interactions are often weak and short-lived. The weakness of the interactions, together with the inevitable coupling of binding and electron transfer processes, makes the application of normal equilibrium methods difficult. One approach to unraveling the relative contributions of redox and binding processes is to model the interactions mathematically, essentially according to Scheme 1:

$$Hb^{3+} + b_5^{2+} \leftrightarrow \{Hb^{3+} ---- b_5^{2+}\} \leftrightarrow \{Hb^{2+} ---- b_5^{3+}\} \longleftrightarrow Hb^{2+} + b_5^{3+}$$

SCHEME 1

where Hb^{3+} and Hb^{2+} and b_5^{3+} and b_5^{2+} represent oxidized and reduced hemoglobin and cytochrome b_5, respectively.

A combination of Brownian dynamic simulation and Marcus theory has allowed the delineation of the contributions of the complex formation and electron transfer steps. The initial binding of oxidized hemoglobin and reduced cytochrome b_5 can be modeled on the basis of Brownian dynamic simulation. This can be implemented using one of a number of publicly available software packages. We have applied the MACRODOX program, made available to us by S. H. Northrup (Department of Chemistry, Tennessee Technological University, Cookeville, TN). We have used this program to calculate the trajectories of cytochrome b_5 and hemoglobin molecules beginning at arbitrary starting positions and identifying whether such trajectories end in complex formation or separation of the two molecules. By repeating these calculations many thousands of times and applying the appropriate statistical arguments it is then possible to predict the nature and characteristics of the binding process between these two molecules.

General Procedure

Protein Complex Formation. Most effort needs to be employed in establishing the structure of the target molecule, in this case hemoglobin. It is necessary to have a reasonably high-resolution structure for the protein in hand in .pdb format. As the computer algorithms are designed to deal with single peptide chains it is necessary to establish pseudo-termini for the four globin subunits. The virtual hemoglobin molecule is then charged, using Tanford–Kirkwood-derived partial charges and a standard amino acid pK look-up table. From the specific charge distribution, the electrostatic field surrounding the hemoglobin molecule is then calculated from the Poisson–Boltzman equation. This process is then repeated for the partner species, the reduced cytochrome b_5 molecule (or any other molecule for which the structure is known and for which the interaction with hemoglobin may be of interest). To obtain meaningful outcomes for the Brownian dynamic simulations, under a particular solution condition, it is necessary to obtain about 100,000 trajectories for the interaction of hemoglobin and cytochrome b_5. We have used an SGI Octane computer, with two processors, to perform the calculations, although any UNIX platform of comparable computing power is appropriate. A single set of calculations using this computer requires approximately 6–12 h of processor time.

The interested reader can obtain a well-written background to this approach from the Web, in the form of a series of notes created by L. F. Ten Eyck and C. Wong (http://mccammon.ucsd.edu/~chem215/index.html); the S. H. Northrup MACRODOX Web site (http://pirn.chem.tntech.edu/macrodox.html) is also recommended for background information. Similar software packages are also available in the form of the UHBD program devised and developed by A. McCammon and colleagues (http://chemcca51.ucsd.edu).

The Brownian dynamics calculations yield some intriguing predictions. The calculations suggest that interaction between all the oxidized hemoglobins and reduced cytochrome b_5 occurs not through the formation of a unique complex, but through an ensemble of structurally related complexes, in which the hemes of each protein make a more or less close contact (Fig. 2). The presence of strong electrostatic fields surrounding all the hemoglobins and cytochrome b_5 serves to enhance complex formation by a factor of 10, due mainly to orientational effects. The outcome of these calculations also yields tables of output containing the relative frequency with which any particular amino acid, on each protein, takes part in binding, together with the structure and energetics of formation of the electron transfer complexes (Table III). It should be noted that this approach treats the binding process as essentially electrostatic in origin. As such, it allows prediction of the impact of such factors as ionic strength, but ignores the roles of other forces in the establishment of the interprotein complex. The validity of such an approach must always be established by comparison with experimental data. These calculations thus yield insight into the

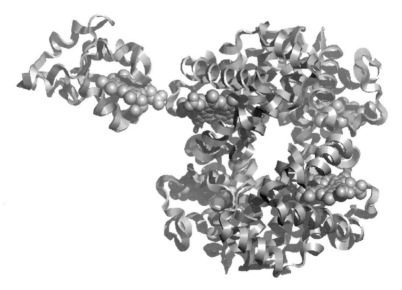

FIG. 2. The structure of a typical encounter complex produced in the reaction of reduced cytochrome b_5 (the smaller molecule) with oxidized human embryonic hemoglobin Gower II ($\alpha_2\varepsilon_2$).

TABLE III
INTERACTION OF CYTOCHROME b_5 AND HEMOGLOBIN SUBUNITS

Subunit	Interaction energy	Theoretical rate	Experimental rate
α	−2.8	1.1×10^4	1.8×10^{4a}
ζ	−0.71	7.9×10^2	8.4×10^{2a}
ε	−4.9	1.7×10^6	2.5×10^{6a}
β	−2.86	1.2×10^4	1.0×10^4
γ	−2.4	8.4×10^3	2.0×10^4

[a] Average of rates from $\alpha_2\beta_2$ and $\alpha_2\varepsilon_2$ and from $\zeta_2\varepsilon_2$ and $\zeta_2\gamma_2$. Interaction energy, kcal/mol; rate constants, $M^{-1} s^{-1}$.

formation and structural characteristics of the electron transfer complex. As this binding process is coupled to electron transfer it is also necessary to model this later process.

Electron Transfer. One powerful approach is to apply Marcus theory. In its simplest formulation this theory predicts that electron transfer is dependent on the distance between the electron donor (the heme group of cytochrome b_5) and the acceptor (the heme group of hemoglobin) and β, a constant that expresses the efficiency of electron tunneling through a particular protein structure. The direct application of this theory to the hemoglobin problem, however, is restricted by the fact that the appropriate β factor is not known and the results of Brownian dynamic simulation show that binding occurs within the context of an ensemble of structures rather than within a unique structure. We have developed a semiempirical approach to this problem. Beratan, Onuchic, and colleagues have developed an analysis that may be used to determine the relative efficiency of electron transfer within a single structure, based on graph theory.[22,23] Thus for a unique structure it is possible to measure the direct distance between redox centers and then determine the relative electron transfer efficiency for the most favorable pathway. This determination can then, in principle, be repeated for the structure of every protein complex predicted from the Brownian dynamic simulation of binding. In practice, however, rather than analyze the many hundreds to thousands of complexes predicted, we find that selection of 50–100 complexes, representing the widest possible range of interredox center distances, can be use to identify "an operative" value for β. Use of this value within the Brownian dynamic simulation then leads to a direct prediction of reaction rates that can be validated by experimental observation.

[22] J. N. Betts, D. N. Beratan, and J. N. Onuchic, *J. Am. Chem. Soc.* **114**, 4043 (1992).
[23] J. J. Ragan and J. N. Onuchic, *Adv. Chem. Phys.* **107**, 497 (1999).

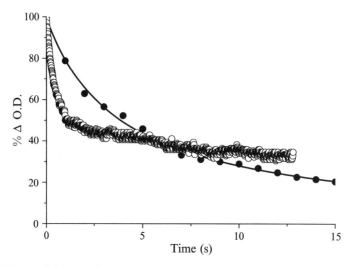

FIG. 3. Stopped-flow reaction time courses for the reaction of 10 μM reduced cytochrome b_5 with human adult hemoglobin (●) and human embryonic hemoglobin Gower II (○). The reaction was performed in 10 mM MES buffer at pH 6.2 in the presence of 100 mM NaCl and 1 mM CO at 25°. The reaction was monitored at 419 nm. The time base for the hemoglobin Gower II ($\alpha_2\varepsilon_2$) reaction is on a 10× shorter time scale.

In our hands these procedures have given predictions of electron transfer rates in excellent agreement with experimental studies of the reaction of hemoglobin with reduced cytochrome b_5 (Fig. 3 and Table III). These findings give strong support to the previous suggestions that the interaction between these two particular proteins is fundamentally electrostatic in origin, although our results indicate reaction via an ensemble of complexes, coupled through a ring of negatively charged amino acids surrounding the heme of cytochrome b_5 and a ring of positively charged amino acids surrounding the heme of hemoglobin, rather than a single complex. The results also indicate that the rate of reduction of any particular globin chain is essentially independent of its partner chain. Interestingly, the rate of reduction within any particular encounter complex is not directly related to the strength of binding within the complex. Significant differences do, however, exist between the rates of reduction for specific types of chain. In particular, the rate of reduction of the embryonic ζ chain is low compared with that of the α chain and undoubtedly reflects the fact that the ζ chain shares only just over 50% homology with the amino acid sequence of the α chain. The now validated procedure should also prove valuable, in future, in identifying the impact of various mutant amino acids on hemoglobin function.

[5] Small-Angle Scattering Techniques for Analyzing Conformational Transitions in Hemocyanins

By HERMANN HARTMANN and HEINZ DECKER

Introduction

The precise delivery of oxygen from respiratory surfaces to the tissues is mediated by cooperative and allosterically regulated carrier proteins such as hemoglobin or hemocyanin. To establish cooperativity these proteins must be able to adopt different conformations. These conformations are characterized by different ligand affinities, which have their basis in different structures as is the case for the deoxy and oxy states of human hemoglobin.

To understand the cooperative interaction of these molecules at the molecular level, the structures of these conformations must be resolved and the transitions between them must be monitored. Because of the nature of sample preparation necessary for various methods, artificial forces may have an impact on the structure. X-ray structures are obtained from crystals. Electron microscopy delivers pictures from biomolecules either fixed on a surface and/or exposed to strong surface forces due to staining or ice buildup. These limitations also hold for newer techniques such as atomic force microscopy (AFM). Two methods are available to study biomolecules in solution under *in vivo* conditions: high-resolution nuclear magnetic resonance (NMR) and small-angle scattering (SAS). NMR delivers structures with atomic resolution from rather small biomolecules with a molecular mass less than 50 kDa. This is not the case for SAS. Here the resolution of the biomolecules obtained by this one-dimensional method is low but sufficient for large biomolecules. There are two methods of SAS—small-angle X-ray scattering (SAXS) and small-angle neutron scattering (SANS)—each of which has its own advantages. SAXS experiments allow better angular resolution, whereas SANS enables the investigation of larger biomolecules. Svergun and Koch[1] nicely illustrated the types of information that can be obtained from SAS data at different angles: shape information at low angles (resolution, >150 Å; $q < 0.5$ Å^{-1}); information relative to the overall fold at midrange angles (resolution, 20–5 Å; q ranges between 0.5 and 1.0 Å^{-1}); and the atomic structure at high angles (resolution, <3.3 Å; $q > 1.5$ Å^{-1}). In practical

[1] D. I. Svergun and M. H. J. Koch, *Curr. Opin. Struct. Biol.* **12**, 654 (2002).

terms, experimental SAS curves with acceptable errors will be obtained in a q range up to 0.5 $\mathring{A}^{-1}$ for larger biomolecules.

Thus, each method to determine structures on its own is ultimately unsatisfying. However, a combination of these methods may give good insight concerning molecular interactions during conformational transitions. Here, we present our methodical approach with SAS to obtain information about the large cooperative hemocyanins and their structural transitions on binding the ligand oxygen or allosteric effectors.

Structure of Hemocyanins

Hemocyanins are respiratory proteins occurring extracellularly in the hemolymph of some arthropods and molluscs. Although the hemocyanins from both stems have a similar oxygen-binding center, they have different structures (Fig. 1).[2,3] Arthropod hemocyanins consist of an integer number of hexamers (1×6, 2×6, 4×6, 6×6, and 8×6) ranging from M_r 450,000 to M_r 3,600,000.[4] Each kidney-shaped subunit (M_r 72,000) folds into three domains. The second domain carries the binding site with its two copper atoms, which bind a molecule oxygen in a side-on coordination. Molluscan hemocyanins are composed of large subunits (M_r ~400,000) folding into seven or eight functional subunits (FUs). Each FU (about 50 kDa) consists of two domains; the N-terminal domain carries the oxygen-binding center, and is almost identical to that from the arthropod hemocyanin. The subunits form the wall of a cylinder. *In vivo* either a decamer or an isologeously dimerized didecamer are formed (Fig. 1). Thus, 160 oxygen-binding sites must be managed.

X-ray structures are known for various hemocyanins: (1) the deoxy state of the spiny lobster hemocyanin *(Panulirus interruptus)*,[5] (2) the oxy and deoxy states of the homohexamer of the horseshoe crab hemocyanin subunit II *(Limulus polyphemus)*,[6,7] and (3) the oxy form of the FU g of the *Octopus* hemocyanin.[8]

[2] K. E. van Holde, K. J. Miller, and H. Decker, *J. Biol. Chem.* **276**, 15563 (2001).

[3] K. E. van Holde and K. I. Miller, *Adv. Protein Chem.* **47**, 1 (1995).

[4] J. Markl and H. Decker, *Adv. Comp. Environ. Physiol.* **13**, 325 (1992).

[5] A. Volbeda and W. G. J. Hol, *J. Mol. Biol.* **209**, 249 (1989).

[6] B. Hazes, K. A. Magnus, C. Bonaventura, J. Bonaventura, Z. Dauter, K. H. Kalk, and W. G. J. Hol, *Protein Sci.* **2**, 597 (1993).

[7] K. A. Magnus, B. Hazes, H. Ton-That, C. Bonaventura, J. Bonaventura, and W. G. J. Hol, *Proteins* **19**, 302 (1994).

[8] M. E. Cuff, K. I. Miller, K. van Holde, and W. A. Hendricksen, *J. Mol. Biol.* **278**, 855 (1998).

Cooperative Oxygen Binding of Hemocyanins

Oxygen binding is highly cooperative, with Hill coefficients of about 7 for molluscan hemocyanins and up to 11 for arthropod hemocyanins (Fig. 1). On the basis of the complex structure of hemocyanin, the Monod–Wyman–Changeux (MWC) model was modified with respect to obvious hierarchies in the structure by assigning their hierarchies of allosteric equilibria.[9,10] This nested MWC model seems to hold for several large protein complexes such as hemocyanins, extracellular hemoglobins, and

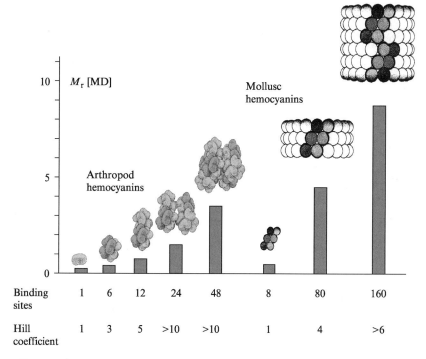

Binding sites	1	6	12	24	48	8	80	160
Hill coefficient	1	3	5	>10	>10	1	4	>6

FIG. 1. Hierarchy in the quaternary structures of hemocyanins from arthropods and molluscs. In arthropods the subunits are kidney shaped. Hexamers or multiples of hexamers (1×6, 2×6, 4×6, and 8×6) are found *in vivo.*. The top trimers are presented in a darker shade. The subunits of mollusc hemocyanins fold into seven or eight functional units (FUs), illustrated as balls. Two of the eight FUs are hidden behind the six other FUs. Differences in primary structure are indicated by different shadings. Cylindric decamers or didecamers are found *in vivo*. The number of oxygen-binding sites and Hill coefficients n_H are given.

[9] H. Decker, C. H. Robert, and S. J. Gill, *in* "Invertebrate Oxygen Carriers" (B. Linzen, ed.), pp. 383–388. Springer, Heidelberg, Germany, 1986.
[10] C. H. Robert, H. Decker, B. Richey, S. J. Gill, and J. Wyman, *Proc. Natl. Acad. Sci. USA* **84,** 1891 (1987).

GroEL.[2,3,9–12] Here, four conformations were predicted. In arthropod hemocyanins the smallest allosteric unit was determined to be the smallest repeating structural unit, which was observed to be the half-molecules. These allosteric units can adopt two conformations, r or t, and, depending on the overall conformation R or T, four conformations can occur for the allosteric unit rT, tT, rR, and tR.

This nesting model requires physical evidence of various conformations that should differ with respect to oxygenation status and effectors bound. This chapter shows how to detect oxygen- and effector-dependent conformations of hemocyanins and how to interpret the conformational transitions between them. For this we applied small-angle X-ray scattering (SAXS) as well as small-angle neutron scattering (SANS).

A Short Guide through the Theory of SAS

In the following discussion, it is assumed that interference between waves scattered at different particles can be neglected. This assumption is approximately correct if either dilute particle solutions are used for the SAS measurements (concentration, $<1\%$) or if the measured intensities are extrapolated to zero concentration.

The scattered intensity of the particle solution is recorded as a function of $q = 4\pi(\sin\theta)/\lambda$, where λ denotes the wavelength and 2θ is the angle between the incident primary beam and scattered radiation. Measurement of the scattering of pure solvent and subtraction from the solution scattering gives the excess scattering $I(q)$ of the particles. The excess scattering is proportional to the squared difference in scattering length density (electron density for X-rays and nuclear spin density for neutrons) between particle and solvent, the so-called excess scattering density. In neutron scattering the contrast between particle and solvent can be varied to provide additional structural information, either by using H_2O–D_2O mixtures or by selective deuteration of the protein.

The distance distribution function $p(r)$ delivers the relative occurrence of the distance r between points of the particle and is given by

$$p(r) = \frac{1}{2\pi^2} \int_0^\infty qr I(q) \sin(qr) dq \tag{1}$$

For determination of the distance distribution from the measured intensities most often the indirect Fourier transformation is used,[13,14] which is based on the inverse transformation of Eq. (1):

[11] N. Hellmann and H. Decker, *Biochem. Biophys. Acta* **1599**, 45 (2002).
[12] N. Hellmann, R. E. Weber, and H. Decker, *Comp. Biochem. Physiol. A Mol. Integr. Physiol.* **136**, 725 (2003).

$$I(q) = 4\pi \int_0^{D_{\max}} p(r) \frac{\sin(qr)}{qr} dr \tag{2}$$

Here $D_{\max}$ denotes the maximum distance occurring in the particle. The radius of gyration can be determined from the Guinier approximation ($q < 1/R_g$):

$$R_g^2 = -3 \frac{d \ln I}{dq^2} \tag{3}$$

which can be obtained graphically from $\ln I(q) = \ln I(0) - R_g^2 q^2/3$, or more accurately from the distance distribution function $p(r)$:

$$R_g^2 = \frac{\int_0^{D_{\max}} p(r) r^2 dr}{2 \int_0^{D_{\max}} p(r) dr} \tag{4}$$

The Debye equation describes the connection between the structure of the particle and the scattering intensity of the particles in vacuum. It is given by

$$I(q) = \sum_{i=1}^{N} \sum_{j=1}^{N} f_i(q) f_j(q) \frac{\sin(qr_{ij})}{qr_{ij}} \tag{5}$$

with the form factor $f(q)$. If all form factors are identical, $f_i = f_j = f$, the equation reduces to

$$I(q) = f^2(q) \sum_{i=1}^{N} \sum_{j=1}^{N} \frac{\sin(qr_{ij})}{qr_{ij}} \tag{6}$$

For a molecular model built from N identical discrete solid spheres (beads), the distance distribution function can be calculated by

$$p(r) = \sum_{i=1}^{N} p_i(r) + 2 \sum_{i=1}^{N-1} \sum_{j=i+1}^{N} p_{ij}(r, r_{ij}) \tag{7}$$

where $p_i(r)$ is the distance distribution function of the sphere and $p_{ij}(r, r_{ij})$ is the cross-term of two spheres with a distance of r_{ij}. Equations for the distribution functions of spheres are given by Glatter.[15]

If the molecule consists of particles that are small compared with the resolution ($f = $ constant), the distance distribution function is given by a sum of δ functions:

[13] O. Glatter, *J. Appl. Crystallogr.* **10**, 415 (1977).
[14] D. I. Svergun, A. V. Semenyuk, and L. A. Feigin, *Acta Crystallogr. A* **44**, 244 (1995).
[15] O. Glatter, *Acta Phys. Austriaca* **52**, 243 (1980).

$$p(r) = \frac{1}{4\pi} \left[\sum_{i=1}^{N} f_i^2 + 2 \sum_{i=1}^{N-1} \sum_{j=i+1}^{N} f_i f_j \delta(r - r_{ij}) \right] \tag{8}$$

To speed up the intensity calculation when Eq. (2) is used to calculate the scattered intensity, the N^2 distances can be sorted in M intervals of width Δr. The number of intervals needed for a precise intensity calculation depends on the resolution, but should be on the order of 1000.[15] Computation time for calculating the N^2 distances is considerably reduced if there are molecular symmetries.

For the discrepancy between experimental data and data calculated from a model we use

$$\chi^2 = \frac{1}{n-1} \sum_{i=1}^{n} \left[\frac{I_{mod}(q_i) - s \cdot I_{exp}(q_i)}{\sigma(q_i)} \right]^2 \tag{9}$$

in the case of intensities and

$$\chi^2 = \frac{1}{n-1} \sum_{i=1}^{n} \left[\frac{p_{mod}(r_i) - s \cdot p_{exp}(r_i)}{\sigma(r_i)} \right]^2 \tag{10}$$

for distance distribution functions. The total number of data points is denoted by n, σ is the error of a data point, and s is a scaling factor.

SAXS Curves of Arthropod Hemocyanins: Simulation and Experiment

Decades ago arthropod hemocyanins were investigated by SAXS by the Kratky group.[16–18] From 1990 to 2000 some crystal structures of arthropod hemocyanins were solved and the SAS techniques became more sophisticated, with new sources such as synchrotron radiation. More recently, low-resolution molecular structures of isolated arthropod hemocyanin subunits and functional units of molluscan hemocyanins were used to obtain a rough "molecular" envelope.[19,20] Using SAXS, Beltramini et al.[21] studied

[16] O. Glatter and O. Kratky, "Small Angle X-ray Scattering." Academic Press, London, 1982.

[17] I. Pilz, K. Goral, M. Hoylaerts, R. Witters, and R. Lontie, Eur. J. Biochem. **105**, 539 (1980).

[18] H. Bernhard, I. Pilz, O. Meisenberger, R. Witters, and R. Lontie, Biochim. Biophys. Acta **748**, 28 (1983).

[19] J. G. Grossmann, S. A. Ali, A. Abbasi, Z. H. Zaidi, S. Steova, W. Voelter, and S. S. Hasnain, Biophys. J. **78**, 977 (2000).

[20] E. Dainese, D. Svergun, M. Beltramini, P. Di Muro, and B. Salvato, Arch. Biochem. Biophys. **373**, 154 (2000).

[21] M. Beltramini, P. Di Muro, R. Favilla, A. La Monaca, P. Mariani, A. L. Sabatucci, B. Salvato, and P. L. Solari, J. Mol. Struct. **475**, 73 (1999).

the temperature stability of an arthropod hemocyanin subunit from *Carcinus estuarii*. The domains seem to be more stable than the interaction between them at higher temperatures.

Simulated SAXS Curves of Known Structures of Arthropod Hemocyanins

Here we present an approach to obtain from known arthropod hemocyanins SAXS curves that will be needed for analyzing SAXS experiments on a molecular basis. We simulate SAXS curves from the atom coordinates of known hexameric structures and try then to fit them to the experimental SAXS curves. For larger hemocyanins, additional information from electron microscopy is used.

Rigid Body Modeling of n × 6-Meric Hemocyanins. (See Fig. 2.) In the case of hierarchically organized arthropod hemocyanins, representative crystal structures are known only for the basic 6-meric unit and for each of the two different classes, crustacean hemocyanins and chelicerate hemocyanins. No significant difference can be deduced from simulated SAXS curves between the 6-meric deoxyhemocyanins from *Panulirus interruptus* and the homohexamer of subunit II from *Limulus polyphemus* in the oxy and deoxy states. In accordance with the common observation that the structures of the hexamers are essentially conserved, the quaternary structures of $n × 6$-meric hemocyanins can be modeled with a small number of free parameters ($3n − 3$ translations and $3n − 3$ rotations) and further reduced by symmetries of the molecule. The arrangement of the hexamers within the $n × 6$-mers is deduced from electron microscopy images.[22–24] The scattering intensity curves reveal two minima for the monomer at about 0.15 and 0.25 Å^{-1}. Whereas the oligomerization to $1 × 6$-mers, $2 × 6$-mers, and $4 × 6$-mers results in an increase in the scattering intensity at $q = 0$ and obvious minima at about $q = 0.075$ Å^{-1}, a shoulder is present at about 0.03 Å^{-1} for the $4 × 6$-mer. The $p(r)$ curves provide information about the size of the protein with $d_{\max}$, which gives the largest distance between any two scattering atoms. The values of 75 Å (monomer), 130 Å (hexamer), 225 Å ($2 × 6$-mer), and 260 Å ($4 × 6$-mer) agree well with roughly estimated values obtained by electron microscopy. In addition, the position of the maxima provides information about the size of the monomers and hexamers. In the case of the $2 × 6$-mer, the maximum at 90 Å coincides with the shoulder of the $p(r)$ curve of the $4 × 6$-mer.

---

[22] F. de Haas, M. M. C. Bijholt, and E. F. J. van Bruggen, *J. Struct. Biol.* **107,** 86 (1991).

[23] F. de Haas and E. F. J. van Bruggen, *J. Mol. Biol.* **237,** 464 (1994).

[24] M. van Heel and P. Dube, *Micron* **25,** 387 (1994).

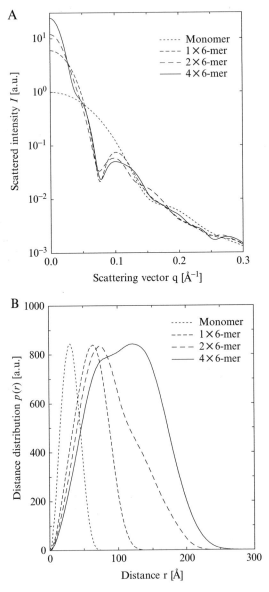

FIG. 2. Simulated SAS curves (A) and $p(r)$ curves (B) of arthropod hemocyanins in different aggregation states. The curves were obtained from the known crystal structures of hexamers. The arrangement between the hexamers is based on parameters deduced from electron microscopy (see text). (A) In all cases the same protein concentration was assumed. (B) The $p(r)$ curves were normalized to the maximum values.

The position of the maximum of the latter at about 140 Å is slightly larger than the shoulder of the 2×6-mer, which gives a rough idea of the distances between the hexamers.

Experimentally Determined SAS Curves

From the atomic model, a distance distribution function $p(r)$ was calculated as described above by counting all distances between the atoms, each weighted by the atomic form factor as shown previously [Eq. (9)]. The agreement of the calculated and experimental distribution functions was compared by χ^2 [Eq. (10)].

Binding of Allosteric Effectors

Allosteric effectors such as L-lactate and urate strongly influence the oxygen-binding behavior of 2×6-meric hemocyanins.[25–27] Thus, it was necessary to prove whether an effector molecule has an influence on the SAXS experiments between q values of 0.015 and 0.04 Å^{-1}. We compared SAXS experiments performed with 2×6-meric lobster hemocyanin from *Homarus americanus* in the presence and absence of L-lactate.[28] The radius of gyration of the molecular models was calculated directly from the coordinates, using the distances between the atoms and the center of mass. Values of 72.1 ± 0.5 Å in the absence of L-lactate and 70.2 ± 0.7 Å in the presence of L-lactate were almost independent of high protein concentration. The decrease in the radius of gyration by about 2 Å in the presence of L-lactate indicates that the hemocyanin molecule becomes more compact. The maximum diameters of lobster hemocyanin were calculated from the $p(r)$ curves. The value of 235 ± 5 Å in the presence of L-lactate was only slightly shorter than that observed in the absence of L-lactate (240 ± 5 Å). This similarity, however, was produced by the observation that the number of large distances in the range between 140 and 240 Å is significantly smaller, compensated by an increase in $p(r)$ at shorter distances.[28] Because the maximum diameter of a hemocyanin hexamer is about 130 Å, as calculated from the X-ray structures,[6] only interhexameric distances can contribute to $p(r)$ above 130 Å. Therefore, we calculated the distance between the centers of mass of the hexamers, $D_{hex-hex}$, from the R_G values because the radius of gyration for the hexamer was known as shown above.

[25] M. Menze, N. Hellmann, H. Decker, and M. K. Grieshaber, *Biochemistry* **39**, 10806 (2000).

[26] H. Hellmann, E. Jaenicke, and H. Decker, *Biophys. Chem.* **90**, 2799 (2001).

[27] B. Zeis, A. Nies, C. R. Bridges, and M. K. Grieshaber, *J. Exp. Biol.* **168**, 93 (1992).

[28] H. Hartmann, B. Lohkamp, N. Hellmann, and H. Decker, *J. Biol. Chem.* **276**, 19954 (2001).

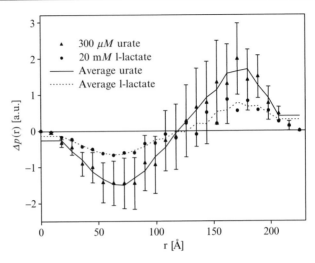

FIG. 3. Difference distance distribution functions $p(r)$ of 2 × 6-meric crustacean hemocyanin from *Astacus leptodactylus* in the presence of allosteric effectors (taken from Gebhardt[29]).

$$D_{\text{hex}-\text{hex}} = 2\sqrt{R_G^2(2 \times 6\text{-mer}) - R_G^2(6\text{-mer})} \qquad (11)$$

From the X-ray structure of the related hexameric hemocyanin from another crustacean, *P. interruptus,* the radius of gyration of 47.7 Å for the R_G(6-mer) was calculated. The distances between the centers of hexamers were then calculated to be 108 Å in the absence of L-lactate and 103 Å in the presence of L-lactate. Thus, the hexamers are shifted toward each other by 5 Å in the presence of L-lactate. This is consistent with the observed decrease of about 5 Å in the maximum diameter as determined above from the distance distribution function.

A similar strategy was applied when the 2 × 6-meric crayfish hemocyanin from *Astacus leptodactylus* binds another effector, urate.[29] Here a similar effect was obtained (Fig. 3) and can be discussed in an analogous way.

Thus, both cases provide the first structural evidence for hemocyanins, that the 2 × 6-mer can adopt two different conformations with different shapes although the hemocyanins are fully oxygenated in all cases.[28,29] This result excludes the simple MWC model for describing the oxygen binding of 2 × 6-meric hemocyanins.

[29] R. Gebhardt, PhD Thesis, University of Mainz, Mainz, Germany, 2000.

Different Conformations of the Oxy and Deoxy States

To analyze conformational transitions of arthropod hemocyanins on oxygenation we choose the structurally well-known 4 × 6-meric hemocyanin from the tarantula *Eurypelma californicum*. SAXS curves were recorded for the oxygenated and deoxygenated states.[30] The scattering experiments were performed with a Kratky camera with a slit collimation system. Scattering intensities were recorded at 113 different angles from $q = 0.01$ to 0.41 Å^{-1}. The radius of gyration is 86.5 ± 0.5 Å for the deoxy form and 88.0 ± 0.5 Å for the oxy form. This indicates that the tarantula hemocyanin is less compact in the oxygenated than in the deoxygenated form. The maximum particle dimension amounts to 25.0 ± 5 Å for the deoxy form and 270 ± 5 Å for the oxy form. A dip in the intramolecular distance distribution function $p(r)$ is more pronounced and shifted to larger distances in the oxy form.[30]

Two different approaches are presented to reveal the underlying mechanism of the conformational switch.

Scattering Spheres Strategy

We constructed a model of the 4 × 6-meric tarantula hemocyanin of identical spherical elements of suitable but arbitrary diameter ("scattering spheres").[31] On the basis of a program of Glatter,[15] scattering curves for such models were calculated and transformed into intraparticle distance distribution functions $p(r)$. These functions were then compared with the experimental distribution functions of the oxygenated and deoxygenated hemocyanin. By trial and error, the arrangement of the spherical elements was modified until the experimental and the theoretical $p(r)$ functions agreed within experimental error. The result is given in Fig. 4A. Each modeled subunit was constructed from 14 scattering spheres, which were arranged in a bilayer forming a typical "kidney-like" shape of an arthropod hemocyanin subunit with the dimensions of known hemocyanins (about 75 × 55 × 60 Å^3). To reconstruct the hexamers, six subunits are arranged as two layers of trimers stacked on each other and rotated against each other around the 3-fold axis by 60°. The four hexamers of tarantula hemocyanin were then arranged according to electron micrographs.[23] Two hexamers are rotated about each other by 90°, forming the 12-meric half-molecules. Two half-molecules are associated in an antiparallel manner and tilted along their length axis by 45°, building up an edge-to-edge contact.

[30] H. Decker, H. Hartmann, R. Sterner, E. Schwarz, and I. Pilz, *FEBS Lett.* **293**, 226 (1996).
[31] R. Sterner, PhD Thesis, University of Munich, Munich, Germany, 1991.

For purposes of fitting, particular scattering spheres were added and removed. The distances between the layers of spheres within the subunits and between the two associated trimers of the hexamers, as well as the distances between the hexamers and between the two 12-meric half-molecules, were changed. In addition, the longitudinal axes of the 12-mers were rotated against each other by different angles. Furthermore, the 12-mers were shifted along these axes by a variable extent. We obtained two models whose distribution functions are in good agreement with the experimental curves for oxy- and deoxyhemocyanin, respectively (Fig. 4A).

Comparison of Models of Oxygenated and Deoxygenated Hemocyanin. A few modifications had to be made regardless of the oxygenation state: the best value for the distance between the longitudinal axes of both dodecameric half-molecules was determined to be 105 Å. Two additional scattering spheres are present at the distal contact area between the hexamers of a dodecamer.

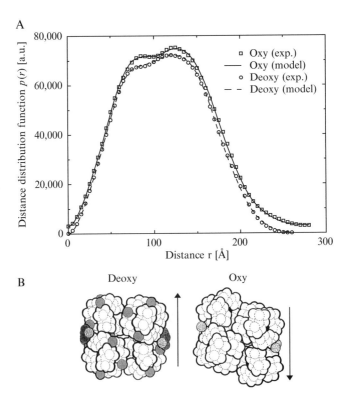

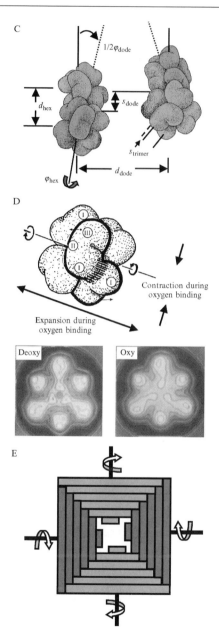

FIG. 4. Comparison of experimentally obtained distance distribution functions $p(r)$ for deoxygenated and oxygenated 4×6-meric tarantula hemocyanin with the fitted $p(r)$. (A) Goodness of fit: SAXS data are taken from Decker et al.[30] The error ranges do not extend beyond the symbols at the middle of the curve. The same goodness of fit was obtained

Level of Hexamers. In the model for the oxy state, the three spacings between the four layers of spheres building up the two trimers within a hexamer are enlarged by about 3.7 Å, compared with the deoxygenated state. Along the 3-fold axis of each hexamer, there is a small gap in the model for the deoxygenated state. To obtain a good fit for the oxy form, the gap had to be filled by two additional spheres, one in the middle of each trimer (Fig. 4B: scattering spheres are black). This shift of mass might indicate that the long solvent channel in the deoxy state is narrowed on oxygenation. Here, domain II carrying the oxygen-binding center should be involved.

Six additional spheres per hexamer must be placed at the outer contact areas of the subunits of a trimer in the deoxy form (Fig. 4B, gray spheres). These spheres must be omitted in the oxy model. Thus, the hexamer is skinnier in the oxy state compared with the deoxy state. The omission of these six scattering spheres per oxygenated hexamer is sensitive to the $p(r)$ function of the oxy model. According to the X-ray structure of *P. interruptus* hemocyanin, these additional spheres are located at the interface between domain III of one subunit and domain I of the neighboring subunit.

---

independent of modeling strategies employed, as described in text. The oxy curve was shifted slightly upward for clarity. (B) Models of deoxygenated and oxygenated 4 × 6-meric tarantula hemocyanin. The subunits and the hexamers within the 24-meric tarantula hemocyanin are arranged as described in text. The modeled subunit consists of 14 scattering spheres arranged in two layers. The radius of a scattering sphere is 13.2 Å. The single subunits are outlined in boldface. Scattering spheres that must be added in the oxygenated state are black, and those that can be detected only in the deoxygenated state are gray (hexameric level) or dotted (interhexameric level). Note that two additional scattering spheres are present at the distal contact area between the hexamers of a dodecamer, independent of the oxygenation state. (C) Free parameters for fitting the 4 × 6-mer with a molecular template (taken from Hartmann and Decker[32]). *Notation:* d_{dode}, distance between the centers of the two 2 × 6-meric half-molecules; d_{hex}, distance between the center of mass of the two hexamers between the 2 × 6-meric half-molecules; s_{dode}, shift of the two 2 × 6-mers along their long rotation axis; d_{hex}, shift of the hexamers with the 2 × 6-meric half-molecules against each other; φ_{hex}, rotation of the hexamers against each other; φ_{dode}, angle between the two 2 × 6-meric half-molecules. (D) *Top:* Proposed mechanism of the conformational transition of the hexamer (after Decker[33]). A conformational transition of the hexamer on oxygenation is proposed. Each subunit of the top trimer is strongly associated with a particular subunit of the bottom trimer by an interaction of domains I (outlined in boldface). A conformational transition leads to a slight counterclockwise twist of the two trimers against each other with respect to the 3-fold axis. To avoid any disconnection between the subunits, they must move closer to the 3-fold axis. This results in a skinnier and longer hexamer on oxygenation. *Bottom:* This transition is confirmed by electron microscopy of negatively stained hexamers.[34] In the deoxy form the central channel could be stained and should therefore be open, but it could not be stained in the oxy form. Thus, the hexamers become longer and skinnier on oxygenation, indicating a pulsation of the hexamer ("breathing"). (E) Scheme of the differential gear model describing the conformational transition between the four hexamers of native tarantula hemocyanin (after Perutz[35]).

Level of Dodecamers. At the contact area between the two hexamers of the dodecameric half-molecules, eight additional scattering spheres (Fig. 4B, black spheres) are present in the deoxy form. According to image analysis of electron micrographs, the C-terminal parts of neighboring subunits from the two closely associated hexamers should be responsible for this "bulge" in the deoxygenated hemocyanin.

Level of 24-mer. When oxygenated, the two dodecamers are shifted along their longitudinal axes by 30 Å compared with the deoxygenated form, and the two half-molecules are tilted against each other by angles up to 35°. According to electron micrographs, the two contact areas between the two 12-meric half-molecules should be involved in this shift in the following way: the top part of the domain III of subunit b in one dodecamer should interact with the top part of the domain III of subunit c in the associated dodecamer, thus forming two hinges for the translational shift of the two 12-mers.

Molecular Modeling Strategy

Another strategy was to fit simulated scattering intensities and $p(r)$ curves of the reconstructed three-dimensional models to the experimental data, based on the molecular structure of 4 × 6-meric tarantula hemocyanin as shown above. As already shown in our work,[32] eight parameters were permitted to adjust.

A grid search procedure was used in the parameter space given by the rotations and translations of the four hexamers. The parameters were independently varied and incremented in each step, typically by 1–5 Å for the translations and by 1–5° for the rotations. A 2-fold symmetry axis was assumed, connecting the two identical 2 × 6-meric half-molecules. Only the C_α atoms of the X-ray structure were used for calculation of the distance distribution functions $p(r)$. About 250,000 model structures were generated in this way and compared with the data. The parameters given in Table I and their errors were derived in the following way. At first a subset of models was selected with χ^2 values no more than 30% higher than the lowest χ^2 value reached in the whole simulation. From these subsets, mean values of the parameters and their standard deviations were calculated.

The comparison revealed (Table I) that all levels of the quaternary structure are involved on oxygenation, the 4 × 6-meric as well as the 2 × 6-meric and 1 × 6-meric structures. The values of the reconstructed 4 × 6-mers obviously agree well with the values obtained experimentally from SAXS experiments. The same calculations were performed on the

[32] H. Hartmann and H. Decker, *Biochim. Biophys. Acta* **1601,** 132 (2002).

TABLE I

FITTED PARAMETERS OF OXY AND DEOXY STATES OF 4×6-MERIC TARANTULA HEMOCYANIN OBTAINED BY SAXS IN COMPARISON WITH THOSE OBTAINED BY ELECTRON MICROSCOPY FROM CLOSELY RELATED 4×6-MERIC HEMOCYANINS[a]

	φ_{hex} (degrees)	φ_{dode} (degrees)	d_{hex} (Å)	d_{dode} (Å)	S_{dode} (Å)	S_{trimer} (Å)	R_g (Å)	d_{max} (Å)
Oxy	105 ± 7	19 ± 4	105.1 ± 1.3	104.4 ± 1.2	18 ± 3	0.5 ± 0.5	88.0 ± 0.5	270 ± 5
Deoxy	131 ± 3	6 ± 3	104.5 ± 0.8	103.5 ± 0.8	4 ± 2	-2.0 ± 0.5	86.5 ± 0.5	250 ± 5
EM	90–120	6–14	102–108	108–119	4–17	—	88–92	

[a] Data from Decker et al.[30] and Hartmann and Decker.[32]

basis of the oxyhomohexamer from *Limulus*, but no significant differences were obtained (data not shown).

Conformational Transition of Basic 6-Meric Hemocyanin. On the basis of these results and the results from our analysis of the "scattering spheres strategy" we suggest the following hypothesis for the conformational transition within a hexamer.[30,32] From the X-ray structures of the homohexamer of *Limulus* hemocyanin it is known that two monomers from the upper and topmost trimers within a hexamer are more closely connected than the monomers within the trimers (the hexamer is more a "trimer of dimers" than a "dimer of trimers") and that a twist of the first domain by 7° with respect to the other two domains was detected by comparing the oxy and deoxy forms with the hexamer of *Panulirus interruptus*.[6,7] This twist of domain I of the three subunits within one trimer in the direction of the other connected trimer can occur only when the two trimers within a hexamer rotate against each other counterclockwise, presented previously as the turning wheel hypothesis[33] (Fig. 4E). In addition, the two trimers are shifted apart by about 2–3 Å according to our analysis (Table I). The subunits must move closer to the 3-fold axis to avoid disruption between subunits, especially with respect to the strong domain I–domain I contacts between the three "tight dimers" of the two trimers. Consequently the central channel will be closed. A comparison with electron micrographs of negatively stained hexamers from spiny lobster obtained at different oxygenation levels confirms our hypothesis (Fig. 4D).[34] In the deoxy form, the central channel was stained and should therefore be open, but it could not be stained in the oxy form. In addition, a movement between domain I and domain III can be deduced from electron micrographs, because staining material is observed only between the domains in the deoxy form. Thus, the hexamers become longer and skinnier on oxygenation, indicating a pulsation of the hexamer ("breathing").

We also propose a scheme for the conformational transition of the four hexamers within 4 × 6-meric hemocyanins (Fig. 4C).[32] Induced by the movement within the hexamers, the two tightly associated hexamers of the 2 × 6-meric half-molecules will also be twisted against each other by $\Delta\varphi_{hex} = 26°$ from 105 to 131°. This rotation will be transferred from one 2 × 6-meric half-molecule to the other one via the two bridges formed by the two heterodimers *b* and *c*. This causes two different movements on

[33] H. Decker, *in* "Structure and Function of Invertebrate Oxygen Carriers" (S. N. Vinogradov and O. H. Kapp, eds.), pp. 89–98. Springer-Verlag, Heidelberg, Germany, 1991.

[34] F. de Haas, J. F. L. van Breemen, E. J. Boekema, W. Keegstra, and E. F. J. van Bruggen, *Ultramicroscopy* **49**, 426 (1993).

oxygenation, an increase in the angle φ_{dode} between the two half-molecules from 6 to 19° and a sliding movement of the two half-molecules along each other from $S_{dode} = 4$ to 18 Å. The consequence is an increase of the longest diagonal d_{max} of the 4 × 6-mer from 250 to 270 Å.

Thus, in tarantula hemocyanin the four cooperative hexameric allosteric units are an embedded 4 × 6-meric structure interacting cooperatively in a two-state manner. Because of the particular structural arrangement between the hexamers, a conformational transition in one hexamer results in a synchronously conformational transition in the other three hexamers to avoid blocking any movement within the 4 × 6-mer. This is in accordance with a previously suggested "differential gear model" by Perutz[35] (Fig. 4E). However, Perutz used this model for the description of the cooperative interactions between the four interacting subunits in the case of tetrameric human hemoglobin, which are noncooperative by themselves.

SANS Experiments with Mollusc Hemocyanin

SAXS experiments have previously been performed with mollusc hemocyanin by Kratky, Pilz, and co-workers[36–42] and then later by others.[19,43] However, all experiments were performed under oxygenated conditions. In addition, for mollusc hemocyanins much less structural information is available compared with arthropod hemocyanins. Preliminary SAXS data concerning native *Octopus vulgaris* hemocyanin in the oxy and deoxy states revealed dramatic conformational changes on oxygenation whereas for *Rapana thomasiana* hemocyanin almost no differences could be detected.[20,44] However, the authors did not discuss their data on a molecular basis.

[35] M. Perutz, "Mechanism of Cooperativity and Allosteric Regulation in Proteins." Cambridge University Press, Cambridge, 1990.
[36] I. Pilz, O. Glatter, and O. Kratky, *Methods Enzymol.* **61,** 148 (1979).
[37] J. Berger, I. Pilz, R. Witters, and R. Lontie, *Eur. J. Biochem.* **80,** 79 (1977).
[38] J. Berger, I. Pilz, R. Witters, and R. Lontie, *Eur. J. Biochem.* **73,** 247 (1977).
[39] I. Pilz, Y. Engelborghs, R. Witters, and R. Lontie, *Eur. J. Biochem.* **42,** 195 (1974).
[40] I. Pilz, K. Walder, and R. Siezen, *Z. Naturforsch. C* **31,** 238 (1976).
[41] O. Kratky and I. A. Pilz, *Q. Rev. Biophys.* **11,** 39 (1978).
[42] J. Berger, I. Pilz, R. Witters, and R. Lontie, *Z. Naturforsch. C* **31,** 238 (1976).
[43] F. Triolo, V. Grauiano, and R. H. Heenan, *J. Mol. Struct.* **383,** 249 (1996).
[44] M. Beltramini, E. Borghi, P. Di Muro, A. La Monica, B. Salvato, and C. Santini, *J. Mol. Struct.* **383** (Suppl. 1), 237 (1996).

SANS Curves of Oxygenated and Deoxygenated KLH1

We investigated the conformational transition of the structurally well-known hemocyanin of the keyhole limpet hemocyanin (KLH) from the marine gastropod *Megathura crenulata,* which is widely used as a hapten carrier, a potent adjuvant immunostimulant, and a therapeutic agent in the treatment of bladder carcinoma.[45] Electron microscopy has revealed similar quaternary structures for the two isoforms KLH1 and KLH2: hollow cylinders with an outer diameter of about 360 Å and two inner rings near both ends of the cylinder. KLH1 forms predominantly didecamers with a height of about 400 Å, consisting of 2 × 10 elongated subunits for which three-dimensional reconstructed models with a resolution of 15 Å have been published.[46]

Figure 5 shows the small-angle intensities of deoxygenated and oxygenated KLH1 as obtained by SANS.[47] The radii of gyration were determined as 163.7 ± 0.5 Å for the oxy form and 165.0 ± 0.6 Å for the deoxy state, indicating that the oxy form of KLH1 is more compact than the deoxy form. But no difference could be determined for the maximal distance D_{max} of about 430 ± 10 Å for both states. The minima and maxima in the scattering curves are less developed for the oxygenated form, and they are shifted to higher q values by 1–2%. In the case of cylindrically shaped molecules deeper minima correspond to a more hollow cylinder whereas shallow minima indicate a more filled cylinder.

SANS-Based Modeling of KLH1

We used two different approaches to build three-dimensional models of KLH1: (1) in the first approach we used geometric forms[47]; (2) in the second approach we used a new Monte Carlo-based algorithm (MCSAS, described below).

Hollow Cylinder Approach On the basis of knowledge from electron microscopy we used (1) hollow cylinders and (2) hollow cylinders with two cylindrical rings to fit the small-angle data, minimizing the discrepancy χ^2 [Eq. (9)].

1. For a simple hollow cylinder the scattered intensity was calculated analytically from three parameters: the length and the outer and inner radii of the cylinder. By a least-squares optimization of these parameters the overall dimensions of KLH1 were determined, but because of the

[45] J. R. Harris and J. Markl, *Micron* **30,** 597 (1999).
[46] E. V. Orlova, P. Dube, J. R. Harris, E. Beckman, F. Zemlin, J. Markl, and M. van Heel, *J. Mol. Biol.* **271,** 417 (1997).
[47] H. Hartmann, A. Bongers, and H. Decker, *Eur. Biophys. J.* **30,** 471 (2001).

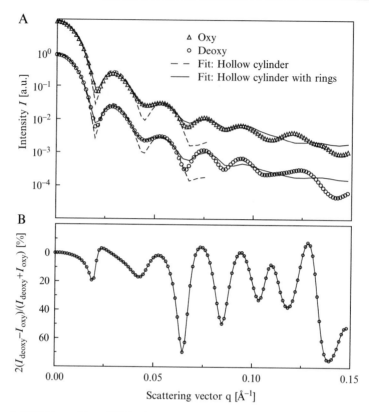

Fig. 5. SANS intensities of KLH1 in the presence and absence of oxygen (from Hartmann et al.[47]). (A) Intensities of oxygenated and deoxygenated KLH1 in D_2O and fitted scattering curves of the cylindrical models. The intensities for deoxygenated KLH1 were divided by a factor of 10 for clarity. (B) Relative difference between the scattered intensities of deoxygenated and oxygenated KLH1.

simplicity of the model, the data could be fitted only in the lowest resolution shells

2. Within the hollow cylinders two additional symmetrically arranged cylindrical rings were added, representing the collars of KLH1. With a least-squares fit the six parameters for the model were determined: the heights of the cylinder and the ring, the outer and inner radius of the cylinder, the inner radius of the ring, and the position of the ring along the cylinder axis. The outer radius of the ring was fixed to the inner radius of the hollow cylinder. When anisotropic scaling parameters such as height and radius of a cylinder were refined, the distribution of scattering points was not considered to be random after the parameters had changed by

more than a few percent from the starting values. In this case a new starting model, with the actual parameters but with a random distribution of points, was generated for the next refinement cycles.

As shown by Hartmann et al.,[47] only minor changes in the quaternary structure of the KLH1 molecules occur on oxygenation. The cylinder becomes slightly longer by about 5 Å and the rings are shifted in the direction of the molecular center by 4 Å, or 10%. In addition, the thickness of the rings decrease from 50 to 46 Å.

Monte Carlo Annealing Algorithm. A well-established method for representing low-resolution structures is the approximation of the molecule by small homogeneous spheres (beads), densely packed in a three-dimensional grid.[48–50] Using the form factor $f(q)$ of a homogeneous sphere the scattered intensity of a model composed of N beads is given by the Debye equation [Eq. (7)]. To avoid artificial density fluctuations the diameter of the spheres (and the distance between neighboring spheres) must be small compared with the experimental resolution, resulting in a typical value of 10^3–10^4 for N in an SAS experiment. The number of parameters is much higher than the information content of an SAS experiment, the degrees of freedom, given by $(q_{max} - q_{min})D_{max}/\pi$.[51] This has several consequences for the modeling procedure: (1) the number of configurations is astronomically large and an exhaustive test against the SAS data is impossible. Therefore an efficient search procedure for possible configurations is needed, like a genetic algorithm[49] or thermal annealing[50]; (2) there is even a large number of models that fit the SAS intensity equally well within the accuracy of the data; and (3) any tiny details and differences between the configurations are below the resolution of the data.

Molecular symmetry can reduce the number of independent spheres substantially, but the regular arrangement of the beads restricts the symmetry to the possible point group symmetries of a three-dimensional lattice. Modeling with densely packed spheres therefore is suited for arthropod hemocyanins, but not for molecules with a C_5 or D_5 symmetry as found for mollusc hemocyanins, because these symmetries are not consistent with the lattice symmetries. Therefore, in addition to regularly packed beads, a model representation using randomly distributed point scatterers[52] was implemented in the MCSAS algorithm. With the random

[48] A. Krebs, J. Lamy, S. N. Vinogradov, and P. Zipper, *Biopolymers* **45**, 289 (1998).
[49] P. Chacon, F. Moran, J. F. Diaz, E. Pantos, and J. M. Andreu, *Biophys. J.* **74**, 2760 (1998).
[50] D. I. Svergun, *Biophys. J.* **76**, 2879 (1999).
[51] L. A. Feigin and D. I. Svergun, "Scattering Analysis by Small-Angle X-Ray and Neutron Scattering." Plenum Press, New York, 1987.
[52] S. J. Henderson, *Biophys. J.* **70**, 1618 (1996).

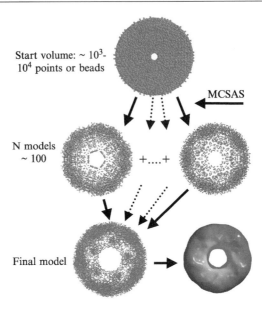

FIG. 6. Scheme of the MCSAS algorithm for *ab initio* determination of molecular models from SAS data (see text).

arrangement, systematic errors in the scattering density at the interface between symmetry-related parts of the molecule can be avoided for all symmetry operations. Statistical density fluctuations introduced in this way are minimized by averaging a large number of molecular models (see below).

In Fig. 6 a scheme shows how the MCSAS algorithm is applied. At the beginning of each pass a starting model is created in the following way: first, an asymmetric unit of a volume enclosing the expected shape of the molecule is filled either with densely packed beads or with randomly distributed scattering points. Then to each point or bead a relative scattering length of 0 or 1 is randomly assigned. Only entities with a scattering length $f = 1$ contribute to the small-angle scattering and to the shape of the molecule. The distance distribution function is calculated with Eq. (7) or Eq. (8) and the scattered intensity is calculated with Eq. (2). Now in a cyclic Monte Carlo procedure new configurations are created. A point or bead is randomly selected, its scattering length is inverted (if $f = 1$ then $f = 0$ and vice versa) and $\Delta\chi^2$ [the change of χ^2; Eq. (9)] is calculated. A criterion is applied to the new configuration similar to the one used by Metropolis *et al.*[53] to allow the algorithm to search over the total configuration space consistent with the experimental data and not to become stuck in local minima. A new configuration with $\Delta\chi^2 \leq 0$ is accepted with probability $P = 1$. For $\Delta\chi^2 > 0$ the probability for accepting the new configuration is given by

$$P = \exp\left(-\Delta\chi^2/\kappa^2\right) \tag{12}$$

where κ is a parameter roughly equal to the mean error of the experimental intensity. During the iteration process κ is adjusted in a way to ensure that the number of iterations to reach equilibration (when χ^2 does not decrease further) is much greater than the number of scattering beads or points. After equilibration the configuration of the molecular model is saved and the algorithm is restarted with a new starting model. Common features of the models are derived by averaging a large number of individual configurations.

Test of MCSAS Algorithm with Homohexameric Hemocyanin

To test the MCSAS algorithm, simulated scattering intensities were calculated from the homohexameric structure of *Limulus* subunit II up to a q value of 0.35 Å^{-1}. A statistical error was assigned to the simulated data, ranging from 1% for $q = 0$ up to 10% for $q = 0.35$ Å^{-1}. For the starting model a cylinder with diameter and height equal to 130 Å was selected and D_3 symmetry was applied. About 100 models were created by the MCSAS program with 10^4–10^5 iterations per model. The averaged SAS model (Fig. 7) compared with the X-ray structure of *Limulus polyphemus* shows the obvious similarity between the two hexamers and how well the atomic structure fits into the modeled hexamer.

Modeling of Oxygenated and Deoxygenated KLH1 with MCSAS Algorithm. For the mollusc hemocyanin KLH1 models were constructed with the MCSAS algorithm for the oxygenated as well as for the deoxygenated forms on the basis of experimental SANS data.[47] The starting volume was restricted to a hollow cylinder of a height less than 400 Å, an outer diameter less than 380 Å, and an inner diameter greater than 50 Å. A D_5 symmetry was applied to the models, thereby reducing the number of independently scattering points by a factor of 10.

Figure 8 shows three-dimensional surface representations of modeled KLH1 molecules in the oxygenated and deoxygenated states. The pictures represent mean models calculated from a total of about 100 MC fits, each with 10^5 iterations. Different shapes of the surface were found for the oxy and deoxy forms. The shape of the outer wall of KLH1 molecules shows

[53] N. Metropolis, A. W. Rosenbluth, N. Rosenbluth, A. H. Teller, and E. Teller, *J. Chem. Phys.* **21,** 1087 (1953).

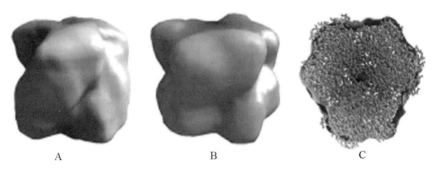

A   B   C

FIG. 7. Comparison of a low-resolution MCSAS model with the X-ray structure. Isodensity representations are shown for the mean MCSAS model (A) and the crystallographic structure (B) of the homohexameric form of *Limulus* subunit II. The modeling was performed with simulated intensity data, calculated from the crystallographic structure. (C) Cross-section perpendicular to the 3-fold axes. The atoms of the X-ray structure are drawn as spheres and the envelope shows the MCSAS model.

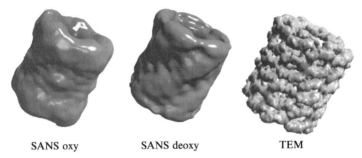

SANS oxy   SANS deoxy   TEM

FIG. 8. SANS model of the KLH1 molecule in the oxy and deoxy states compared with a reconstruction from transmission electron microscopy (TEM). Each of the SANS models shown has been calculated from an average of 100 individual MCSAS models. The data used for calculating the isodensity surface of the TEM model have been kindly provided by Orlova *et al.*[46]

swelling bulges orientated parallel to the cylindrical axis in the oxy form. In the deoxy form they are twisted along this axis. In both oxygenation states masses were concentrated at the two ends of the cylindrical molecules, forming the internal "collars," but no masses were observed in the middle of cylinders. The radii of gyration for the models of deoxygenated and oxygenated KLH1 molecules were calculated as 164.8 and 163.6 Å, in good agreement with the experimental values of 165.0 and 163.7 Å, respectively. In comparing the 15-Å resolution structure of KLH1 based on electron microscopy with our oxy and deoxy models, the transmission electron microscopy (TEM) model seems to be more similar to the deoxy structure.

To reveal differences between the modeled structures with respect to the length and radius of the cylindrical KLH1 molecules, the radius of gyration was split in a component R_z along the axes of the cylinder and a radial component R_{xy}. Then R_z is given as

$$R_z = \mathrm{sqrt}\left(\sum_{i=1}^{N} f_i z_i^2 / \sum_{i=1}^{N} f_i\right) \tag{13}$$

and the weighted radius R_{xy} as

$$R_{xy} = \mathrm{sqrt}\left(\sum_{i=1}^{N} f_i(x_i^2 + y_i^2) / \sum_{i=1}^{N} f_i\right) \tag{14}$$

x_i, y_i, z_i, and f_i are the coordinates and scattering lengths of the N scattering points. For the deoxygenated molecule we determined the values to $R_z = 108$ Å and $R_{xy} = 125$ Å and for the oxygenated molecule to $R_z = 109$ Å and $R_{xy} = 122$ Å. Thus, the mass-weighted length of the modeled oxy-KLH1 molecule is greater than that of the deoxygenated form, whereas the mean diameter is smaller.

Comparison of SAS Experiments and Electron Microscopy. A comparison of reconstructed hemocyanins based on small-angle scattering and TEM reveals that the latter gives dimensions that are larger than those obtained from SANS or SAXS. This discrepancy between the two methods, small-angle scattering and TEM, is often observed for large extracellular respiratory proteins such as hemoglobin from *Lumbricus terrestris*[48] and *Macrobdella decora*[54] and hemocyanins from *Helix pomatia*[55,56] and *Eurypelma californicum*.[30] In all cases the size of the molecules determined by electron microscopy was larger than that determined by SAXS. This is surprising because one should expect an inverse correlation. Whereas SAXS and SANS take into account the water molecules within the hydration shell around the protein, this is not the case in TEM, using negatively stained specimens.[16,57] According to Parak *et al.*,[58] the thickness

[54] O. H. Kapp, M. Scmick, I. Pilz, and S. N. Vinogradov, *in* "Invertebrate Dioxygen Carriers" (G. Preaux and R. Lontie, eds.), pp. 219–223. Leuven University Press, Leuven, Belgium, 1990.

[55] J. F. L. van Bremen, J. H. Ploegman, and E. F. J. van Bruggen, *Eur. J. Biochem.* **100,** 61 (1979).

[56] I. Pilz, O. Kratky, and I. Moring-Claesson, *Z. Naturforsch. B* **25,** 600 (1970).

[57] D. I. Svergun, S. Richard, M. H. J. Koch, Z. Sayers, S. Kuprin, and G. Zaccai, *Proc. Natl. Acad. Sci. USA* **95,** 2267 (1998).

of the hydration shell should be about 2.5 Å. Thus, dimensions obtained by SAXS and SANS should be slightly larger than those determined by electron microscopy, not smaller.

For arthropod hemocyanins these differences can be corrected. In the case of the 2 × 6-meric lobster hemocyanin the two hexamers,[28] and in the case of the 4 × 6-meric tarantula hemocyanin the half-molecules (2 × 12-mers),[30] must be shifted together by about 1 nm to yield contacts between the half-molecules and to be in reasonable agreement with the SAXS data. The reason may be that the hemocyanin molecules are slightly distorted by the preparation of the negatively stained specimen, resulting in larger differences between the two loosely connected half-molecules. Such an artifact cannot arise when SAS is applied, because untreated proteins are investigated in solution. Thus, SAS, which analyzes molecules in solution, may be considered an important check for the three-dimensional reconstruction of macromolecules based on negatively stained specimens by TEM. This seems to be important especially for proteins with a large and complex quaternary structure.

Acknowledgments

We thank R. May, A. Bongers, R. Gebhardt, G. Görigk, B. Lohkamp, T. Nawroth, I. Pilz, and R. Sterner for help in performing the SAS experiments, and J. R. Harris and J. Markl for helpful criticism. This work was supported by the Deutsche Forschungsgemeinschaft, the BMBF, the ILL (Grenoble), the HASYLAB (Hamburg), the Center for Science and Medical Research (NMFZ Mainz), and the Center for Material Science (MWFZ Mainz).

---

[58] F. Parak, H. Hartmann, M. Schmidt, G. Corongiu, and E. Clementi, *Eur. Biophys. J.* **21**, 313 (1992).

[6]   Multivalent Protein–Carbohydrate Interactions: Isothermal Titration Microcalorimetry Studies

By TARUN K. DAM and C. FRED BREWER

Introduction

A wide variety of cellular and pathological processes are mediated by carbohydrate–protein interactions.[1,2] These interactions generally require high-affinity binding. However, carbohydrate-binding proteins (lectins) typically show low affinities for simple mono- and oligosaccharides. Higher affinity interactions occur when lectins, which are oligomeric proteins, bind to the carbohydrate chains of cell surface glycolipids and glycoproteins, which possess multiple binding epitopes. As a consequence, considerable attention has been given toward understanding the underlying mechanisms responsible for the enhanced affinity of multivalent carbohydrates for lectins.

Insight into the thermodynamic basis for the enhanced affinities of multivalent (clustered) glycosides binding to the lectins concanavalin A (ConA) and *Dioclea grandiflora* (DGL) has been obtained by isothermal titration microcalorimetry (ITC). ITC measurements provide direct determinations of binding enthalpy, ΔH, the association constant, K_a, and the number of binding sites of the protein, n. From measurements of K_a, the free energy of binding, ΔG, can be calculated. The entropy of binding, ΔS, is obtained from ΔH and ΔG. Thus, ITC measurements can determine the complete thermodynamics of binding of a carbohydrate to a lectin.

ConA and DGL are mannose/glucose-specific lectins with similar binding specificities. They possess relatively high affinities for the monovalent trisaccharide 3,6-di-O-(α-D-mannopyranosyl)-α-D-mannopyranoside (mannotriose; **4** in Fig. 2) as compared with mannose. Synthetic multivalent clustered glycosides bearing multiple terminal mannose or mannotriose residues show increased affinities for ConA and DGL up to nearly 100-fold as assessed by enzyme-linked lectin assay[3,4] and hemagglutination inhibition.[5] To gain insight into the thermodynamic basis for the enhanced

[1] A. Varki, *Glycobiology* **3**, 97 (1993).
[2] H. Lis and N. Sharon, *Chem. Rev.* **98**, 637 (1998).
[3] R. Roy, D. Page, S. F. Perez, and V. V. Bencomo, *Glycoconj. J.* **15**, 251 (1998).
[4] D. Page and R. Roy, *Glycoconj. J.* **14**, 345 (1997).
[5] T. K. Dam, R. Roy, D. Pagé, and C. F. Brewer, *Biochemistry* **41**, 1351 (2002).

affinities of these multivalent saccharides, binding of synthetic dimeric analogs of α-D-mannopyranoside (Fig. 1) and di-, tri-, and tetrameric
analogs of 3,6-di-*O*-(α-D-mannopyranosyl)-α-D-mannopyranoside (Fig. 2)
to ConA and DGL was studied by ITC. The results show that ITC yields
important thermodynamic insight into the mechanism(s) of enhanced affinities of these multivalent carbohydrates for these two lectins. The results
also show that the negative cooperativity that occurs during binding is
associated with the multivalent carbohydrates and not the proteins, and
that entropy effects play a dominant role in the enhanced affinities of these
ligands.

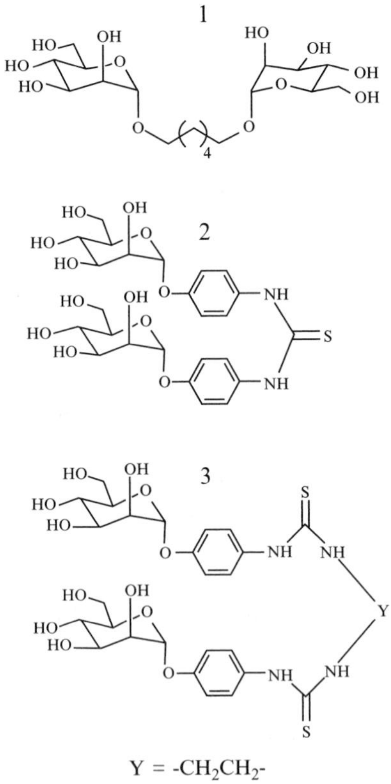

Y = -CH₂CH₂-

FIG. 1. Structures of bivalent analogs **1–3**.

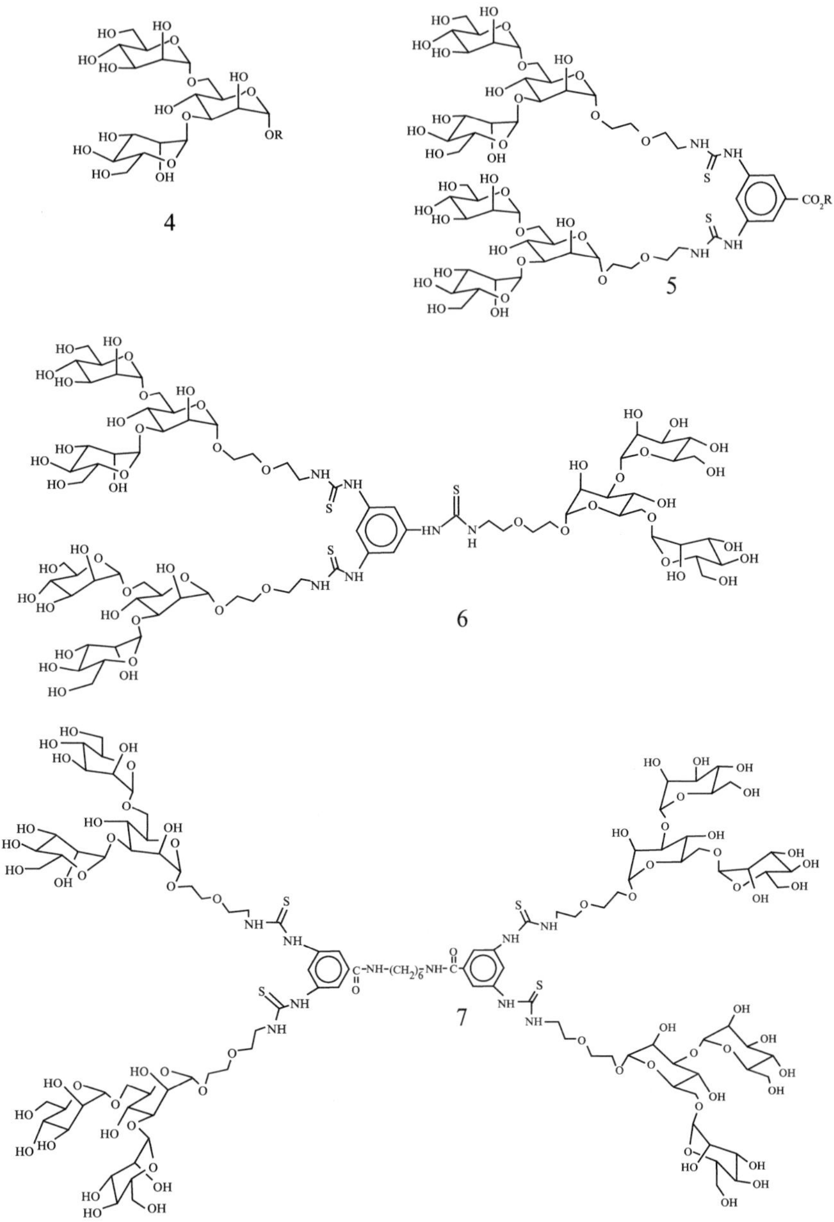

FIG. 2. Structures of monovalent mannotriose (**4**) and its multivalent analogs **5–7**.

Materials and Methods

DGL was isolated from *Dioclea grandiflora* seeds obtained from North Eastern Brazil (Albano Ferreira Martin, Sao Paulo, Brazil) as previously described.[6] The concentration of DGL was determined spectrophotometrically at 280 nm, using $A^{1\%, 1\ cm} = 12.0$ at pH 5.2, and expressed in terms of monomer (M_r 25,000).[6] ConA was purchased from Sigma (St. Louis, MO) or was prepared from jack bean *(Canavalia ensiformis)* seeds (Sigma) according to the method of Agrawal and Goldstein.[7] The concentration of ConA was determined spectrophotometrically at 280 nm, using $A^{1\%, 1\ cm} = 12.4$ at pH 5.2,[8] and expressed in terms of monomer (M_r 25,600).

Methyl α-D-mannopyranoside (MeαMan), *p*-aminophenyl α-D-mannopyranoside, *p*-nitrophenyl α-D-mannopyranoside, and methyl 3,6-di-*O*-(α-D-mannopyranosyl)-α-D-mannopyranoside (mannotriose, **4**), were purchased from Sigma. The synthesis of carbohydrate analogs **2** and **3**[4] and of **5**, **6** and **7** has previously been reported.[3] Synthesis of **1** will be reported elsewhere. The concentrations of the carbohydrates were determined by modification of the Dubois phenol-sulfuric acid method using appropriate monosaccharides as standards.[9]

Isothermal Titration Calorimetry

ITC experiments with multivalent carbohydrates and lectins were performed with an MCS instrument from Microcal (Northampton, MA). Injections of 4 μl of carbohydrate solution were added from a computer-controlled 250-μl or 100-μl microsyringe at intervals of 4 min into the sample solution of lectin (cell volume, 1.3424 ml) with stirring at 350 rpm. Control experiments performed by making identical injections of saccharide into a cell containing buffer without protein showed insignificant heats of dilution. The experimental data were fitted to a theoretical titration curve, using software supplied by Microcal, with ΔH (enthalpy change, kcal/mol), K_a (association constant, M^{-1}), and n (number of binding sites per monomer) as adjustable parameters. Unlike some common practices, the n value was never set to a predecided value; rather, it was kept as a variable parameter. The lectin concentration used was in terms of its monomer. The quantity $c = K_a M_t(0)$, where $M_t(0)$ is the initial macromolecule

[6] R. A. Moreira, A. C. H. Barros, J. C. Stewart, and A. Pusztai, *Planta* **158,** 63 (1983).
[7] B. B. L. Agrawal and I. J. Goldstein, *Biochim. Biophys. Acta* **147,** 262 (1967).
[8] I. J. Goldstein and R. D. Poretz, *in* "The Lectins" (I. E. Liener, N. Sharon, and I. J. Goldstein, eds.), p. 35. Academic Press, New York, 1986.
[9] S. K. Saha and C. F. Brewer, *Carbohydr. Res.* **254,** 157 (1994).

concentration, is of importance in titration calorimetry.[10] All experiments were performed with $1 < c < 200$. Thermodynamic parameters were calculated from the equation

$$\Delta G = \Delta H - T\Delta S = -RT \ln K_a$$

where ΔG, ΔH, and ΔS are the changes in free energy, enthalpy, and entropy of binding. T is the absolute temperature and $R = 1.98$ cal mol^{-1} K^{-1}.[11]

Data Fitting and Concentrations of Lectins and Carbohydrates

ITC[12] and crystallographic studies[13,14] have established a single carbohydrate-binding site for mannotriose per subunit of ConA and DGL. Therefore, ITC binding data for the multivalent sugars for the two lectins were fitted using a single-site model with a lectin concentration based on their respective subunit molecular weights. Because there is no binding cooperativity between subunits of ConA or DGL (as demonstrated by ITC with monovalent sugars), each monomer of the lectin was treated as a separate binding entity.

Importantly, the concentrations of the carbohydrate analogs used in the ITC experiments are molar units, not equivalent units of epitopes, because the units in the thermodynamic binding equations are molar.

Experimental Conditions for ITC Measurements with Multivalent Sugars

At pH 7.2 and NaCl concentrations greater than 0.15 M when ConA and DGL are tetramers. Multivalent carbohydrate analogs (Figs. 1 and 2) are observed to bind and precipitate with both proteins at lectin concentrations between 25 and 60 μM and at nearly stoichiometric ratios of the sugars. This is due to the cross-linking activities of the multivalent carbohydrates and the oligomeric structures of ConA and DGL. The precipitation reactions are inhibited by the presence of monovalent MeαMan or dissolved on addition of the monosaccharide. ITC measurements, however, require a nonprecipitating solution during the titration experiment. This

[10] T. Wiseman, S. Williston, J. F. Brandt, and L.-N. Lin, *Anal. Biochem.* **179**, 131 (1989).
[11] T. K. Dam, R. Roy, S. K. Das, S. Oscarson, and C. F. Brewer, *J. Biol. Chem.* **275**, 14223 (2000).
[12] T. K. Dam, S. Oscarson, and C. F. Brewer, *J. Biol. Chem.* **273**, 32812 (1998).
[13] J. H. Naismith and R. A. Field, *J. Biol. Chem.* **271**, 972 (1996).
[14] D. A. Rozwarski, B. M. Swami, C. F. Brewer, and J. C. Sacchettini, *J. Biol. Chem.* **273**, 32818 (1998).

was accomplished by adjusting the pH, ionic strength, and concentration of the lectins as described below.

The dimer–tetramer equilibrium of ConA is sensitive to pH[15,16] and NaCl (T. K. Dam and C. F. Brewer, unpublished observations, 2002). Therefore, ITC experiments were performed at low salt and pH (5.0), conditions under which the lectins are predominantly dimeric. A dimeric lectin does not form precipitates on binding a bivalent sugar. The formation of precipitates with tetravalent analogs is also slower if the concentrations of the ligand and lectin are relatively low. Taking advantage of the higher affinities of the multivalent carbohydrates, ITC experiments at relatively low concentrations were performed. In particular, measurements with 5–7 were performed with <20 μM lectins and 150–600 μM carbohydrates, which are significantly lower concentrations compared with those generally used in lectin–carbohydrate titrations. Time-dependent precipitation of the lectins occurred with certain ligands after the run. Thus, the ITC experiments with multivalent ligands were performed under conditions in which formation of insoluble complexes was arrested or considerably slowed. The resulting quality of the data and fitting was excellent (cf. Fig. 3).

Scatchard and Hill Plot Analyses of ITC Data

In addition to determining thermodynamic binding parameters, raw ITC data for binding of multivalent carbohydrates to lectins can be used to construct Scatchard and Hill plots to determine whether cooperativity effects occur on binding. In turn, this can provide important insights into the binding mechanisms of these molecules. Calculations from the ITC data of the parameters for such plots are described below.

The total concentration of ligand $X_t(i)$ as well as of lectin $M_t(i)$ after the ith injection and the heat evolved on the ith injection, $Q(i)$, are readily available from the ITC raw data file. The concentration correction is automatically done by Origin software from Microcal.

The concentration of bound ligand $X_b(i)$ after the ith injection is

$$X_b(i) = [Q(i)/(\Delta H \times V_0)] + X_b(i-1) \tag{1}$$

where $Q(i)$ (μcal) is the heat evolved on the ith injection, ΔH (cal mol^{-1}) is the enthalpy change, V_0 (ml) is the active cell volume, and X_b (mM) is the concentration of bound ligand. X_b is equal to M_b, the concentration of bound protein, and, in the present study of multivalent ligands, the more

[15] G. H. McKenzie, W. H. Sawyer, and L. W. Nichol, *Biochim. Biophys. Acta* **263**, 283 (1972).
[16] M. Huet, *Eur. J. Biochem.* **59**, 627 (1975).

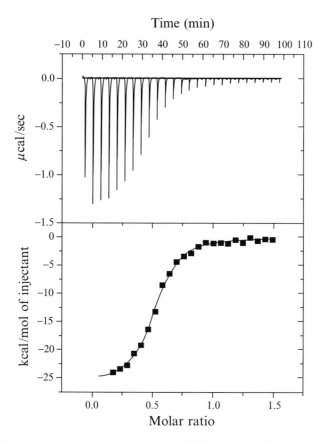

FIG. 3. Isothermal titration microcalorimetry (ITC) profile of ConA (0.020 mM) with bivalent analog **5** (0.39 mM) at 27°. *Top:* Data obtained for 30 automatic injections, 4 μl each, of **5**. *Bottom:* Integrated curve showing experimental points (■) and the best fit (—). The buffer was 0.1 M sodium acetate at pH 5.2 with 0.1 M NaCl and CaCl$_2$ and MnCl$_2$ (5 mM each).

general expression is $M_b = X_b \times$ (functional valency of ligand). The concentration of free ligand (X_f) after the ith injection is determined as follows:

$$X_f(i) = X_t(i) - X_b(i) \tag{2}$$

For Scatchard analysis, $r(i)$ was plotted against $r(i)/X_f(i)$, where $r(i)$ is $X_b(i) \times$ (functional valency of ligand)$/M_t(i)$, and Hill plots were constructed by plotting $\log[Y(i)/1 - Y(i)]$ versus $\log[X_f(i)]$, where $Y(i)$ is

$X_b(i) \times$ (functional valency of ligand)/$M_t(i)$, which are modified versions of the Scatchard and Hill plots[17] that take into account the functional valency of the ligand (i.e., valency of the ligand determined by ITC and not its structural valency). Hill plots were disposed around the zero point on the ordinate as observed for monovalent mannotriose, only after multiplication with the functional valencies of the respective carbohydrates.

A program was created, using Microsoft Excel, for Scatchard and Hill plots. Work sheet data are copied from the ITC raw data file and placed in appropriate columns of the program. After calculation the program shows the profiles of Scatchard and Hill plots. DeltaGraph (RockWare, Golden, CO) is used for further analysis of the plots.[5]

The validity of the information obtained from the Hill plot [log(Y/1-Y) versus log(X_f)] is tested by directly fitting the binding data of monovalent mannotriose to the Hill equation. The slope of the Hill plot is observed to be the same by direct fitting or by plotting the Hill equation data. Attempts at directly fitting the ITC data for the multivalent analogs fail because the Hill slope values change throughout the binding process as a result of cooperativity effects (below).

The raw ITC data including those of control experiments were taken from experiments performed at low and comparable lectin concentrations. Low-concentration titrations inhibit precipitation and self-association of the molecules involved during the experiments. As a consequence, fittings of the ITC data were excellent, which, in turn, significantly reduced the error margins in the raw data. Use of identical experimental conditions and comparable concentrations of lectins and the carbohydrates in all experiments allowed a valid comparison of the data and the profiles. If the raw data are taken from an ITC experiment with poor fitting, Scatchard and Hill plot profiles remain largely unreliable because of the significant error margin. The conclusions made from the present Scatchard and Hill plot analyses were experimentally confirmed by reverse ITC experiments (see below).

Reverse Isothermal Titration Calorimetry

Reverse ITC experiments were performed to determine the microscopic binding parameters of individual epitopes of a multivalent analog. In a reverse ITC experiment, titration was performed as described above with the following exception: in individual titrations, injections of 4 μl of ConA solution were added from the computer-controlled 100-μl microsyringe at intervals of 4 min into a cell containing carbohydrate solution (cell volume, 1.358 ml) dissolved in the same buffer as the lectin, while stirring

---

[17] E. Di Cera, "Thermodynamic Theory of Site-Specific Binding Processes in Biological Macromolecules." Cambridge University Press, New York, 1995.

at 350 rpm. Data were fitted in a one- or two-site model depending on the functional valency of the carbohydrate. Control experiments performed by making identical injections of ConA into a cell containing buffer with no carbohydrate showed insignificant heats of dilution.[18]

Results and Discussion

Determination of the Enhanced Affinities of Multivalent Carbohydrates by ITC

ITC experiments show that the multivalent carbohydrates in Figs. 2 and 3 possess higher affinities for ConA and DGL than their respective monovalent analogs.[11] For example, **1** has a 4-fold higher K_a for ConA relative to MeαMan, and a 20-fold higher affinity for DGL (Table I). Analogs **2** and **3** show 4- to 5-fold higher K_a values for ConA (Table I) relative to *p*-aminophenyl α-D-mannopyranoside, and 2- to 4-fold higher K_a values for DGL (Table I). Analogs **5**, **6**, and **7** show 6-, 11-, and 35-fold higher K_a values for ConA, respectively, and 5-, 8-, and 53-fold higher K_a values, respectively, for DGL relative to the mannotriose (Table I). These data are in agreement with hemagglutination inhibition data.[5]

ITC-Determined n *Values Are Inversely Proportional to the Functional Valency of Multivalent Carbohydrates*

ITC data indicate that the number of binding sites (n) per monomer of ConA[19] and DGL[12] for simple mono- and oligosaccharides such as Meα-Man and mannotriose is close to 1.0. These values agree with the X-ray crystal data for a single binding site on each monomer of ConA[13,20] and DGL[14] and confirm that MeαMan and mannotriose (**4**) are monovalent ligands for ConA and DGL.

The n values (Table I). for **1**, **2**, **3**, and **5** binding to ConA and DGL are $\sim$0.5. Analogs **1**, **2**, and **3** contain two mannose residues in each molecule, whereas **5** possesses two mannotriose residues (Figs. 1 and 2, respectively). Therefore, all four analogs are structurally bivalent. The theoretical value of n for binding of a bivalent carbohydrate to ConA and DGL is 1.0/2 = 0.5, because each ligand molecule contains two binding epitopes and can occupy two protein-binding sites for these two lectins. Hence, the ITC n

---

[18] T. K. Dam, R. Roy, D. Pagé, and C. F. Brewer, *Biochemistry* **41**, 1359 (2002).
[19] D. K. Mandal, N. Kishore, and C. F. Brewer, *Biochemistry* **33**, 1149 (1994).
[20] J. H. Naismith, C. Emmerich, J. Habash, S. J. Harrop, J. R. Helliwell, W. N. Hunter, J. Raftery, A. J. Kalb-Gilboa, and J. Yariv, *Acta Crystallogr. D Biol. Crystallogr.* **50**, 847 (1994).

TABLE I

THERMODYNAMIC BINDING PARAMETERS FOR CONCANAVALIN A AND DIOCLEA GRANDIFLORA LECTIN WITH MULTIVALENT SUGARS AT 27° [a]

	K_a $(M^{-1} \times 10^{-4})$	$-\Delta G$ (kcal/mol)	$-\Delta H$ (kcal/mol)	$-T\Delta S$ (kcal/mol)	n (no. of sites per monomer)
Concanavalin A					
MeαMan[b]	1.2	5.6	8.4	2.8	1.0
1	5.3	6.5	15.2	8.7	0.54
p-APMan[c]	1.3	5.6	7.8	2.2	1.0
2	4.7	6.4	17.0	10.6	0.52
3	5.4	6.5	16.6	10.1	0.52
4[d]	39	7.6	14.7	7.1	1.0
5	250	8.7	26.2	17.5	0.53
6	420	9.0	29.0	20.0	0.51
7	1350	9.7	53.0	43.3	0.26
Dioclea grandiflora lectin					
MeαMan[b]	0.46	4.9	8.2	3.3	1.0
1	10.6	6.8	14.8	8.0	0.56
p-APMan[c]	0.7	5.2	7.3	2.1	1.0
2	1.6	5.7	14.3	8.6	0.60
3	2.5	6.0	14.8	8.8	0.57
4[d]	122	8.3	16.2	7.9	1.0
5	590	9.2	27.5	18.3	0.51
6	1000	9.6	32.2	22.6	0.40
7	6500	10.6	58.7	48.1	0.25

[a] Errors in K_a range from 1 to 7% for ConA and from 7 to 10% for DGL; errors in ΔG are less than 1% for ConA and 1% for DGL; errors in ΔH are 1–4% for ConA and 1–7% for DGL; errors in $T\Delta S$ are 1–7% for ConA and 1–2% for DGL; errors in n are less than 2% for ConA and less than 1% for DGL.
[b] Methyl α-D-mannopyranoside.
[c] p-Aminophenyl α-D-mannopyranoside.
[d] Methyl 3,6-di-O-(α-D-mannopyranosyl)-α-D-mannopyranoside.

values for **1**, **2**, **3**, and **5** confirm that these analogs are functionally bivalent for the two lectins.

The ITC-derived n values for **7** are 0.26 for ConA and 0.25 for DGL (Table I). Analog **7** is structurally tetravalent because it possesses four mannotriose residues. The n values for **7** are consistent with the theoretical value of $n = 1.0/4 = 0.25$ for a tetravalent carbohydrate binding to either lectin. This indicates that all four mannotriose moieties of **7** bind to ConA and DGL.

These results demonstrate that the functional valency of the multivalent carbohydrates in Figs. 1 and 2 is inversely proportional to their

ITC-derived n value for ConA and DGL, two lectins with well-defined single binding sites per monomer. The results also demonstrate that the structural valency (number of same epitopes in an analog) and functional valency (number of epitopes that bind) of analogs **1**, **2**, **3**, **5**, and **7** are the same.

Functional Valency of a Multivalent Analog Can Differ from Its Structural Valency

The ITC-derived value of n is 0.51 for binding of **6** to ConA (Table I). This value of n differs from the predicted value of 0.33 based on the structural valency of this triantennary analog. Thus, **6** is functionally bivalent in binding to ConA as indicated by ITC measurements. On the other hand, the n value for **6** binding to DGL is 0.40, which is between 0.50 (bivalent binding) and 0.33 (trivalent binding). The n value of 0.40 suggests that greater than half of the molecules of **6** are involved in trivalent binding to DGL and less than half are involved in bivalent binding. Thus, ITC data indicate that analog **6** is bivalent in binding to ConA, but that it is bivalent and trivalent in binding to DGL. The reason for this difference in functional valency of **6** for the two lectins is unknown. This difference in functional valency of **6** for ConA and DGL is also interesting because the structures of the two lectins are similar.[14] These results indicate the importance of ITC n values in determining the relationship between the structural and functional valency of a multivalent carbohydrate for specific lectins.

ΔH Increases in Proportion to Valency of High-Affinity Multivalent Carbohydrates

ITC data obtained for binding of multivalent analogs with relatively high affinities show that the observed ΔH value per mole of analog is approximately the sum of the ΔH values of the individual epitopes. For example, bivalent analogs **1**, **2**, and **3** binding to ConA (Table I) have ΔH values of -15.2, -17.0, and -16.6 kcal/mol, respectively, which are almost two times greater than the ΔH values of the respective monovalent analogs MeαMan ($\Delta H = -8.4$ kcal/mol) and p-aminophenyl-α-D-mannopyranoside ($\Delta H = -7.8$ kcal/mol). Similar data exist with DGL (Table I).

The ΔH value of bivalent analog **5** is -26.2 kcal/mol for binding to ConA, which is almost twice the ΔH value of -14.7 kcal/mol for the monovalent analog mannotriose (Table I). The same is true for **5** binding to DGL (Table I). The ΔH value for tetravalent analog **7** binding to ConA is -53.0 kcal/mol, which is approximately four times as great as the ΔH value of monovalent mannotriose (-14.7 kcal/mol). The ΔH value for **7** binding to DGL is -58.7 kcal/mol, which is also nearly four times as great as the ΔH value of the free mannotriose (-16.2 kcal/mol). The ΔH

values of analog **6** are more complicated because the analog is functionally bivalent for ConA ($n = 0.51$), and partially trivalent for DGL ($n = 0.40$) (Table I). The results in Table I thus provide evidence that the ΔH values of relatively high-affinity multivalent carbohydrates binding to ConA and DGL are approximately the sum of the ΔH values of the individual binding epitopes in the analogs.

Similar observations have been made for the binding of a trivalent system of receptor and ligand derived from vancomycin and D-Ala-D-Ala.[21]

TΔS *Does Not Directly Increase in Proportion to Valency of Multivalent Carbohydrate Analogs*

Although ΔH scales proportionally to the number of binding epitopes in high-affinity multivalent carbohydrates, $T\Delta S$ does not. Instead, $T\Delta S$ is much more negative than if it proportionally scaled to the number epitopes in the molecule. For example, **7** contains four trimannosyl-binding epitopes, and to a first approximation the ΔH value for ConA of -53 kcal/mol is four times the ΔH of -14.7 kcal/mol for mannotriose (Table I). However, the observed $T\Delta S$ value for **7** is -43.3 kcal/mol, not -28.4 kcal/mol, if it scaled with the $T\Delta S$ value of -7.1 kcal/mol for mannotriose (Table I). The resulting ΔG value of **7** would also be much greater if $T\Delta S$ scaled with valency because the difference between ΔH and $T\Delta S$ would be greater. However, the observed ΔG value(s) for **7** are much smaller. The same is true for the other multivalent carbohydrates in Table I. Hence, the data in Table I indicate that ΔH values scale with ligand valency, but that the $T\Delta S$ values do not, resulting in the comparatively reduced enhancement of K_a (ΔG) values that is characteristic of the binding of a multivalent ligand to separate receptor molecules in solution. In the present case, a single molecule of **7** binds to four separate ConA or DGL molecules in solution (Fig. 4B). The distance between two carbohydrate epitopes of the multivalent analog in the present study is not great enough to span separate binding sites in ConA or DGL. This is also consistent with the cross-linking activities of these carbohydrates.

These thermodynamic results are different from those of a multivalent ligand binding to a single receptor molecule possessing multiple binding sites (Fig. 4A). In such an instance, the increase in affinity would be much greater than those observed in Table I. For example, binding of a triantennary complex carbohydrate to the hepatic asialoglycoprotein receptor results in a $\sim 10^9$ M^{-1} inhibition constant relative to the $\sim 10^3$ M^{-1} value

[21] J. Rao, J. Lahiri, L. Isaacs, R. M. Weis, and G. M. Whitesides, *Science* **280,** 708 (1998).

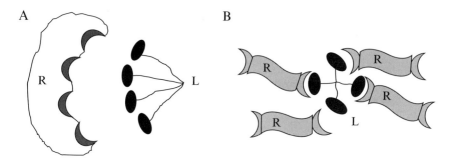

FIG. 4. Schematic representation of two different models of multivalent interactions. Location of the binding sites on the receptor molecule and the length and flexibility of the linker region of the ligand influence the way multivalent binding takes place. For example, epitopes of a multivalent ligand (L) can bind to various sites of a single receptor molecule (R), as shown in (A), leading to significant enhancement of affinity. On the other hand, different epitopes of a multivalent ligand (L) can bind [in (B)] to separate receptor molecule (R) as the epitopes cannot span the two binding sites of the same receptor molecule. Affinity enhancement is relatively modest in this case. The latter model (B) represents the binding interactions described in the present study.

of the corresponding monovalent oligosaccharide.[22] Even more dramatic is the increase in affinity to $\sim 10^{17}\ M^{-1}$ of a trivalent derivative of vancomycin binding to a trivalent derivative of D-Ala-D-Ala, in which the affinity of the corresponding monovalent analogs is $\sim 10^6\ M^{-1}$.[21] In the latter study, thermodynamic measurements showed that both ΔH and $T\Delta S$ scaled proportionally to the number of binding epitopes in the molecules. As a result, the resulting ΔG value is comparatively much greater (more negative) than those observed in the present study (Table I).

Thermodynamic Basis for Enhanced Affinities of Multivalent Analogs

The enhanced affinities of the multivalent carbohydrates in Figs. 1 and 2 for ConA and DGL are associated with their epitopes binding to separate lectin molecules.[11] The observed K_a values for **7**, for example, are the average of the four microscopic K_a values at each of its four epitopes, because each of the four epitopes is involved in binding to a separate lectin molecule. It follows that if ΔH is constant at each epitope and is approximately the same as for mannotriose (Table I),[11] then increases in the overall microscopic K_a values of the four epitopes require more favorable $T\Delta S$

[22] Y. C. Lee, R. R. Townsend, M. R. Hardy, J. Lonngren, J. Arnarp, M. Haraldsson, and H. Lonn, *J. Biol. Chem.* **258**, 199 (1983).

contributions compared with mannotriose. The fact that multivalent analogs must have microscopic K_a values associated with their multiple epitopes leads to a prediction regarding their interactions with lectins such as ConA and DGL. Sequential binding of each epitope in the multivalent analogs must result in diminished valency at every step of binding. Therefore, sequential binding of the epitopes of the analogs should occur with decreased affinity (negative cooperativity) as their effective valency decreases. This predicted decrease in functional valency of the multivalent analogs and concomitant negative binding cooperativity can be demonstrated with Scatchard and Hill plots of the raw ITC binding data for each analog, and by reverse ITC experiments (see below).

Negative Cooperativity in Binding of Multivalent Carbohydrates

Scatchard and Hill plots are widely used to detect positive or negative cooperativity in the binding of monovalent ligands to multisubunit proteins. Ligand binding without cooperativity gives rise to linear Scatchard and Hill plots, with a slope of 1.0 for the latter. Ligand binding with positive cooperativity gives rise to Hill plots with slopes greater than 1.0, whereas ligand binding with negative cooperativity gives Hill plots with slopes less than 1.0. Thus, Hill plots have the advantage of assigning numerical values to the degree of binding cooperativity, as compared with Scatchard plots.[23] ITC binding data for mannotriose (4) and 5–7 were used to construct both Scatchard and Hill plots for data analysis.

In using Scatchard and Hill plot analysis of the binding of a multivalent ligand such as the carbohydrates in the present study, the term for the fraction of bound ligand, X_b/M_t, is corrected for the valency of the sugar to give $(X_b) \times$ (functional valency of sugar)$/M_t$, which is a modification to accommodate multivalent ligands in the classic Hill and Scatchard plot analysis.

Scatchard plots of the ITC data for mannotriose binding to ConA and DGL are linear (Fig. 5) and agree with a previous report.[19] Linear Scatchard plots for the binding of monosaccharides to ConA have also been reported.[24] Scatchard plots generated with the binding data of analogs 5–7 are curvilinear for both ConA and DGL. A representative plot is shown for analog 7 binding to ConA (Fig. 5). The concave nature of the Scatchard plots as seen in Fig. 5 suggests that multivalent analogs 5–7 bind to both lectins with negative cooperativity. To confirm this observation, Hill plots of the ITC binding data for 5–7 to both lectins were examined.

[23] G. Scatchard, *Ann. N.Y. Acad. Sci.* **51**, 660 (1949).
[24] L. L. So and I. J. Goldstein, *Biochim. Biophys. Acta* **165**, 398 (1968).

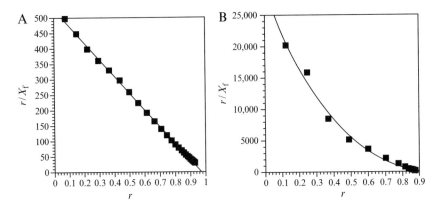

FIG. 5. Scatchard plots of the ITC raw data for (A) monovalent **4** (0.8 m*M*) and (B) multivalent **7** (0.24 m*M*) binding to ConA (0.020 m*M*).

The Hill plot of ITC data for mannotriose (**4**) binding to ConA is essentially a straight line (Fig. 6) with a slope of 0.94, which is close to the value of 1.0 for noncooperative binding interactions.[17,25] A similar plot with a slope of 0.95 is observed for DGL.

The absence of allosteric interactions in ConA and DGL on binding mannotriose allows application of Hill plot analysis to the ITC binding data for analogs **5–7** because the incremental heats measured on sugar addition and binding are proportional to the number of moles of ligand bound, and are not due to allosteric transitions in the lectins. All the ITC binding data for multivalent sugars **5–7** were obtained at sugar concentrations comparable to that of monovalent mannotriose used in the control experiments.

Hill plots of the ITC data for analogs **5–7** binding to ConA are curvilinear rather than linear. The plots are disposed around the zero point on the ordinate, as observed for monovalent mannotriose (**4**) (Fig. 6), only after the corrections for the respective functional valencies of the multivalent sugars have been made. This provides further confirmation regarding the validity of the ITC-derived functional valencies of **5**, **6**, and **7** for ConA and DGL as described above. The tangent slopes of three progressive point intervals of the x axis of the Hill plots were determined in order to assess the changing slope values of the curved plots. All three analogs possess initial tangent slope values between 0.8 and 0.9, which gradually decrease to 0.35, 0.24, and 0.22 for **5**, **6**, and **7**, respectively. These decreasing values of tangent slopes are the signature of increasing negative cooperativity in the

[25] L. Stryer, "Biochemistry." W. H. Freedman and Company, New York, 1988.

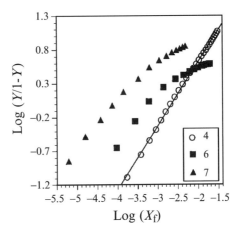

FIG. 6. Hill plots of ITC raw data for monovalent **4** (0.8 mM) and multivalent **6** (0.67 mM) and **7** (0.24 mM) binding to ConA (0.020 mM).

binding of analogs **5–7** to ConA. Similar observations were also made with the same multivalent analogs and DGL. However, the tangent slopes of the curvilinear Hill plots for DGL are distinct from these for ConA.

The physical basis for the increasing negative binding cooperativity of **5–7** can be rationalized, in part, by reduction in the functional valency of the analogs as they bind an increasing number of lectin molecules. For example, Fig. 7 shows the various microequilibrium constants for **7** as its four epitopes sequentially bind one, two, three, and four molecules of ConA (or DGL). The functional valency of unbound **7** (species A) is four, the functional valency of **7** with one bound lectin molecule (species B) is three, the functional valency of **7** with two bound lectin molecules (species C) is two, and the functional valency of **7** with three bound lectin molecules (species D) is one. Sequential occupancy of the four epitopes of the analog would result in a gradual decrease in valency and binding affinity. Increasing negative cooperativity as documented by the curvilinear Hill plots is consistent with the decreasing binding affinity.

Another physical factor that may play a role in the curvilinear Hill plots of **5–7** with ConA and DGL is the formation of noncovalent cross-linked complexes between lectin molecules and multivalent carbohydrates.[11] Figure 7 is overly simplified in that each lectin molecule, represented as a monomer in the scheme, is actually a dimer under the conditions of the experiment (pH 5.2 and low ionic strength). Hence, each lectin molecule (ConA or DGL) is capable of binding and cross-linking the multivalent carbohydrates in the present study. Analogs **5** and **6** are divalent for ConA

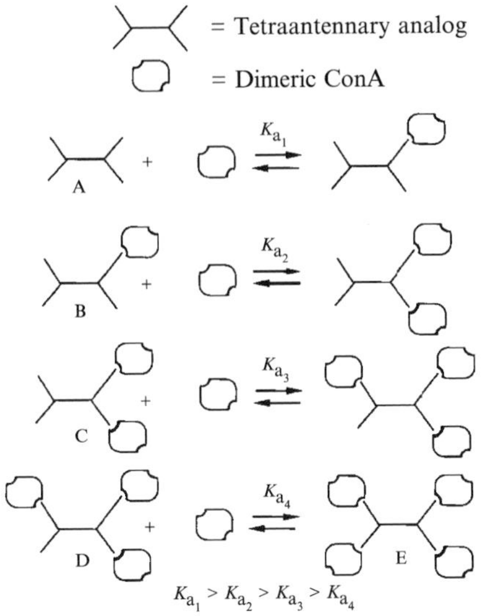

= Tetraantennary analog

= Dimeric ConA

A + $\rightleftharpoons$ K_{a_1}

B + $\rightleftharpoons$ K_{a_2}

C + $\rightleftharpoons$ K_{a_3}

D + $\rightleftharpoons$ K_{a_4} E

$K_{a_1} > K_{a_2} > K_{a_3} > K_{a_4}$

FIG. 7. Four microequilibrium constants of the tetravalent analog **7** can be represented by K_{a_1}, K_{a_2}, K_{a_3}, and K_{a_4}, for binding of a dimeric ConA molecule to the first arm of **7** (species A), to the second arm of **7** (species B), and so on. Hence, the observed (macroscopic) ΔG values of **7** (ΔG_{obs}) for ConA are the average of the four microscopic ΔG terms, or $\Delta G_{obs} = (\Delta G_1 + \Delta G_2 + \Delta G_3 + \Delta G_4)/4$. The relative values of ΔG_1, ΔG_2, ΔG_3, and ΔG_4 must decrease on the basis of the decreasing valencies of A, B, C, and D (which have the same valencies as tetra-, tri-, bi-, and monovalent analogs). Thus, it is expected that $K_{a_1} > K_{a_2} > K_{a_3} > K_{a_4}$ for **7** binding to ConA as shown. An increasing level of cross-linking with the progression of binding will also contribute to the decreasing microscopic binding constants.

(Table I), and may form linear cross-linked complexes with the protein. Analog **7** is tetravalent for ConA (Table I) and also expected to form non-covalent cross-linked complexes with the lectin. Hence, it is possible that the formation of cross-linked lattices with analogs **5**–**7** may effect their binding interactions with successive lectin molecules, which contributes to the increasing negative cooperativity observed in their Hill plots with both lectins. Preliminary evidence indicates that such cross-linking interactions along with the decreasing functional valencies of **5**–**7** on sequential binding of lectin molecules play a role in the curvilinear Hill plots of the analogs.

Microscopic K_a Values Associated with Multivalent Carbohydrates

The observed macroscopic K_a values for multivalent analogs **5**–**7** in Table I are the average of the microscopic binding free energy terms $(-\Delta G)$ of the different epitopes of the analogs. Analog **7**, for example, possesses four mannotriose epitopes, and the ITC data in Table I indicate that all four epitopes are involved in binding to separate molecules of ConA and DGL, respectively. As shown in Fig. 7, the four microequilibrium constants for the four epitopes of 7 can be represented by K_{a_1}, K_{a_2}, K_{a_3}, and K_{a_4}, for binding of a ConA or DGL molecule to the first epitope of 7 (tetravalent species A), to the second epitope of 7 (trivalent species B), to the third epitope of 7 (bivalent species C), and to the fourth epitope of 7 (monovalent species A), respectively. Hence, the observed macroscopic ΔG values of 7 (ΔG_{obs}) for ConA and DGL in Table I are the average of the four microscopic ΔG terms, or ΔG_{obs} (7) = ($\Delta G_1 + \Delta G_2 + \Delta G_3 + \Delta G_4$)/4. In valency terms, the species A, B, C, and D are structurally similar to 7, 6, 5, and 4 (mannotriose), respectively. As the ΔG values decrease consistently from 7 to 4 (7 > 6 > 5 > 4) (Table I), the relative values of ΔG_1, ΔG_2, ΔG_3, and ΔG_4 obtained with A, B, C, and D, respectively, must also decrease in the same order. Therefore the microscopic association constants for 7 binding to ConA and DGL would gradually decrease ($K_{a_1} > K_{a_2} > K_{a_3} > K_{a_4}$).

The following section provides direct ITC experimental evidence for the decreasing microscopic K_a values of 5 and 6 in binding ConA molecules to their different epitopes.

Determination of Microscopic Thermodynamic Binding Parameters of Epitopes of Multivalent Carbohydrates by Reverse ITC

Reverse ITC experiments were performed with ConA titrated into solutions of 5 and 6 and 4 (mannotriose).[18] Data were fitted in a one- or two-site model, depending on the functional valency of the carbohydrate, which was determined from the "normal" ITC data as presented in Table I. Importantly, selection of other values for the number of epitopes failed to provide a fit of the data.

ConA was titrated into a solution of monovalent mannotriose (4) as a control experiment and fitted with a one-site model. The ITC profile is shown in Fig. 8 and the data are shown in Table II. The n value for 4 in the reverse experiment (Table II) is 0.99, which agrees well with the n value of 1.0 for the normal titration (Table I). The K_a value for 4 in the reverse titration (Table II) is $6.3 \times 10^5 \ M^{-1}$ as compared with $3.9 \times 10^5 \ M^{-1}$ for the normal titration (Table I). The ΔH value for 4 in the reverse experiment is -13.1 kcal/mol (Table II) as compared with -14.7 kcal/mol for the normal titration (Table I). Therefore, the results of the reverse ITC of ConA with monovalent mannotriose agree with its previously reported "normal" ITC results.

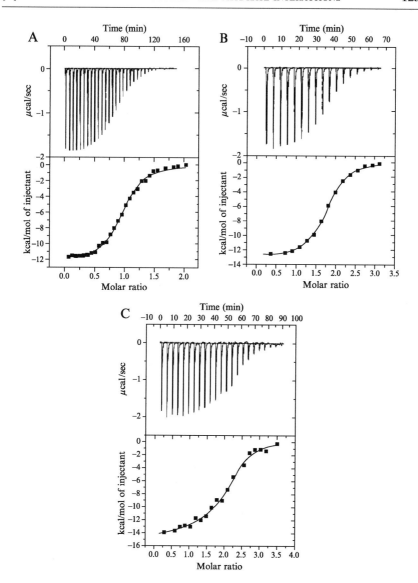

Fig. 8. Profiles of reverse ITC experiments in which ConA was injected from a syringe into cells containing solutions of **4**, **5**, and **6**. In (A) 1 mM ConA was titrated into a solution of 50 μM mannotriose (**4**); in (B) 970 μM ConA was titrated into a solution of 16 μM carbohydrate analog **5**; and in (C) 1 mM ConA was titrated into a solution of 20 μM carbohydrate analog **6** at 27°. The buffer was 0.1 M sodium acetate buffer, 150 mM sodium chloride, pH 5.2. The fitting models were selected according to the functional valencies of the carbohydrates.

TABLE II

Reverse Isothermal Calorimetry-Derived Thermodynamic Binding Parameters for
ConA with Mannotriose and Multivalent Sugar Analogs 2 and 3 at 27° [a]

	K_{a_1} $(M^{-1} \times 10^{-5})$	ΔK_{a_2} $(M^{-1} \times 10^{-5})$	$-\Delta G_1$ (kcal/mol)	$-\Delta G_2$ (kcal/mol)	n_1 (no. of sites per monomer)	n_2
Mannotriose[b]	6.2	—	7.9	—	0.99	—
5	161	8.8	9.8	8.1	0.97	0.94
6	460	8.6	10.4	8.1	1.05	1.09

	$-\Delta H_1$ (kcal/mol)	$-\Delta H_2$ (kcal/mol)	$-T\Delta S_1$ (kcal/mol)	$-T\Delta S_2$ (kcal/mol)
Mannotriose[b]	13.1	—	5.2	—
5	12.5	12.3	2.7	4.2
6	13.3	12.2	2.9	4.1

[a] Errors in K_a are less than 7%; errors in ΔG are less than 5%; errors in n are less than 4%; errors in ΔH are less than 4%; errors in $T\Delta S$ are less than 7%.
[b] Methyl 3,6-di-O-(α-D-mannopyranosyl)-α-D-mannopyranoside.

The reverse ITC profile of ConA with 5 is shown in Fig. 8. The data were fitted using a two-site model because data in Table I from the normal ITC experiment demonstrated two binding sites for 5. The data from the reverse ITC experiment are given in Table II and show individual thermodynamic data for the two binding epitopes of 5 with ConA. The n value for the first binding epitope (n_1) of 5 is 0.97, and n for the second epitope (n_2) is 0.94. This indicates that both mannotriose epitopes of 5 are fully bound to ConA.

Analog 6 is functionally bivalent with ConA, as in 5. Importantly, the reverse ITC data of ConA with 6 (Fig. 8) could be fit only with a two-binding site model, not with a three-site model, consistent with our previous findings.[11] The values of n for the two sites are 1.05 (n_1) and 1.09 (n_2), which suggest that two of the three epitopes of 6 are involved in binding with ConA.

The reverse ITC measurements show two microscopic K_a (K_{a_1} and K_{a_2}) of two epitopes of 5 (Table II). K_{a_1} is $1.6 \times 10^7 \ M^{-1}$ and K_{a_2} is $8.8 \times 10^5 \ M^{-1}$. Hence, the microscopic affinity constant of the first epitope of 5 is 18 times greater than that of the second epitope. The latter value, in turn, is close to the affinity constant of mannotriose (4). The observed macroscopic ΔG value for 5 binding to ConA in the normal ITC experiment is -8.7 kcal/mol. Table II shows that ΔG_1 is -9.8 kcal/mol and ΔG_2 is -8.1 kcal/mol for the two epitopes of 5. The average value of ΔG_1 and ΔG_2 is -9.0 kcal/mol, which is similar to the macroscopic ΔG

of -8.7 kcal/mol observed in Table I. Thus, the observed macroscopic K_a for **5** in Table I agrees with the average of the two microscopic K_a values for **5** reported in Table II. The microscopic association constants of the two sites of **6** are K_{a_1}, $4.6 \times 10^7 \ M^{-1}$; and K_{a_2}, $8.6 \times 10^5 \ M^{-1}$ (Table II). Thus, there is a 53-fold higher affinity of the first binding site of **6** relative to its second binding site for ConA. Interestingly, the affinity of the second site on **6** is similar to that of **5**, whereas the affinity of the first site on **6** is 2.5-fold greater than that of the first site of **5**. This may relate to the greater structural valency of **6** relative to **5**, although their functional valencies are the same.

Table I shows that the observed macroscopic ΔG value for **6** binding to ConA in the normal ITC experiment is -9.0 kcal/mol. Table II shows that the microscopic ΔG_1 is -10.4 kcal/mol and ΔG_2 is -8.1 kcal/mol for the two epitopes of **6** that bind two molecules of ConA. The average of ΔG_1 and ΔG_2 is -9.3 kcal/mol, which is similar to the observed macroscopic ΔG value of -9.0 kcal/mol for **6** in Table I. Thus, the observed macroscopic K_a for **6** in Table I agrees with the average of the two microscopic K_a values for **6** reported in Table II. This is an important means of confirming the validity of the two microscopic K_a values determined in the reverse ITC measurement.

Therefore, these data show that binding of the first mannotriose epitope of a bivalent analog (such as **5** and **6**) to ConA occurs with much higher affinity than binding of the second epitope to another molecule of ConA. These results are consistent with the progressively decreasing K_a values as shown in Fig. 7 and the increasing negative cooperativity as demonstrated by Scatchard and Hill plot analysis.

The microscopic enthalpy of binding of the two epitopes (ΔH_1 and ΔH_2) of **5** are essentially the same, which in turn are similar to that for mannotriose in Table II (-13.1 kcal/mol). The same is also true for analog **6**. These findings agree with the conclusion reached in our study, using normal ITC measurements,[11] that the two functional binding epitopes of **5** and **6** possess essentially equal microscopic ΔH values and are additive in the observed macroscopic ΔH value (Table I).

Microscopic entropy of binding ($T\Delta S$) values of the two epitopes ($T\Delta S_1$ and $T\Delta S_2$) of **5** and **6** were calculated from corresponding microscopic ΔG (ΔG_1 and ΔG_2) and microscopic ΔH (ΔH_1 and ΔH_2) values of **5** and **6**, respectively. As shown in Table II, entropy of binding for the first epitope of **5** is 1.5 kcal/mol more favorable relative to its second epitope. Similarly, there is a 1.2 kcal/mol more favorable entropy of binding value for the first epitope of **6** compared with its second epitope. These results provide a direct demonstration of the favorable entropy effects in the enhancement of affinity of bivalent carbohydrates.

Conclusions

Insight into the thermodynamics of multivalent lectin–carbohydrate interactions has been obtained by ITC. Binding data obtained with the lectins ConA and DGL and several mono- and multivalent carbohydrate ligands show that ITC can be used to determine their thermodynamic binding parameters. The enhanced K_a values and functional valencies of the multivalent analogs for the two lectins were directly obtained by ITC. In the present studies, the increase in affinities of multivalent carbohydrates for ConA and DGL are due to relatively favorable contributions of $T\Delta S$ to ΔG. Negative cooperativity also occurs, which is shown to be due to the decreasing functional valencies of the multivalent carbohydrates as well as to their lectin cross-linking activities. Reverse ITC experiments with two functionally bivalent ligands directly determined the microscopic K_a, ΔH, and n values of the two epitopes in each carbohydrate molecule. The results demonstrated that the higher affinity of the first epitope in each molecule is due to a more favorable contribution of $T\Delta S$ and not ΔH. In terms of designing multivalent carbohydrate inhibitors for lectins, the use of ITC is shown to be a powerful method for obtaining insight into the mechanism(s) of the enhanced affinities of these ligands.

Acknowledgments

This work was supported by Grant CA-16054 from the National Cancer Institute, Department of Health, Education and Welfare, and by Core Grant P30 CA-13330 from the same agency (C.F.B.).

[7] Calorimetric Analysis of Mutagenic Effects on Protein–Ligand Interactions

By Frederick P. Schwarz

Introduction

Application of microcalorimetry to the investigation of protein–ligand interactions in solution has increased dramatically, particularly in the use of isothermal titration calorimetry (ITC). In an ITC experiment, the power exchanged, >10 nW between a protein solution (1.5 ml) and a reference buffer solution (1.5 ml), is monitored in an adiabatic enclosure as 1- to 20-μL aliquots of a ligand solution are added to the protein solution up to and beyond saturation of the protein-binding sites. Conceptually, the power or heat exchanged per each addition of titrant to the solution is a measure of the change in internal energy or enthalpy of the protein solution on binding of the ligand (the binding enthalpy $= \Delta_b H^\circ$) and the rate of change in the amount of exchanged heat per each addition of the titrant as the protein becomes saturated in the bound state is a measure of the binding constant for the interaction, K_b. Accordingly, the analysis of the measurements in terms of a binding model is straightforward, as shown in the analysis below, and is based on the conservation of mass.[1] Because the measurements are performed under thermodynamic equilibrium conditions, the binding constant yields the free energy change for the interaction, ΔG°, through

$$\Delta_b G^\circ = -RT \ln K_b \qquad (1)$$

where T is the absolute temperature of the system and R is the ideal constant, $8.315 \ \mathrm{JK^{-1} \ mol^{-1}}$. The entropy change for the protein solution on ligand binding is the binding entropy, $\Delta_b S^\circ$, and through the fundamental equation of thermodynamics,

$$\Delta_b G^\circ = \Delta_b H^\circ - T\Delta_b S^\circ \qquad (2)$$

The dependence of $\Delta_b H^\circ$ on temperature yields the heat capacity change accompanying the binding interaction, $\Delta_b C^\circ$. This thermodynamic information is useful in quantifying on a macroscopic level whether a binding interaction is driven by an increase in entropy or a decrease in enthalpy of the solution. Just how these thermodynamic changes, which are measured on the macroscopic level, can be described in terms of a binding mechanism on the molecular level necessitates the acquisition of structural information on the protein and ligated protein from such structural methods as X-ray crystallography, nuclear magnetic resonance (NMR), and small-angle neutron scattering (SANS). Then how are these thermodynamic changes correlated with the structural changes induced on the molecular level through mutagenesis? Because changes in the free energy, enthalpy, and entropy in a system are usually treated as a sum of contributions from each component of a system, one approach that has been successful is to correlate these thermodynamic binding quantities with changes in the solvent-accessible surface area of the polar and nonpolar amino acid residues in the protein that accompany the binding interaction.[2,3]

[1] T. Wiseman, S. Williston, J. F. Brandts, and L. N. Lin, *Anal. Biochem.* **179,** 131 (1989).
[2] I. Luque, O. L. Mayorga, and E. Freire, *Biochemistry* **35,** 13681 (1996).
[3] E. J. Sundberg, M. Urrita, B. C. Braden, J. Isern, D. Tsuchiya, B. A. Fields, E. L. Malchiodi, J. Tormo, F. P. Schwarz, and R. A. Mariuzza, *Biochemistry* **39,** 15375 (2000).

Relationship between Mutagenesis and Thermodynamic Changes

Mutagenesis of proteins involves replacement of the amino acid side chains with side chains of different size and/or different polarity and, thus, alteration of the solvent-accessible surface area of both the polar and non-polar amino acid residues. Mutagenesis at the site where the binding interaction occurs (direct mutagenesis) would be expected to induce changes in the binding thermodynamics. Thermodynamic analysis of the binding of the hen egg white lysozyme (HEL) antigen to the mutated binding site on the Fv fragment of the D1.3 antibody to be described is an example of how to analyze interactions involving direct mutagenesis. Alternatively, mutations at sites not located at the binding interaction site (indirect muta-genesis) induce changes in the binding thermodynamics through structural changes in the protein that affect the topography of the binding site. Thermodynamic analysis of the cooperative binding of cyclic nucleoside monophosphate (cNMP) ligands to the amino-terminal domain of cAMP receptor protein (CRP) with a mutation along the protein subunit interface to be described is an example of how to analyze reactions involving indirect mutagenesis. In both types of mutagenesis, structural data play an important role in the analysis of the effect of mutagenesis on binding reactions on the molecular level.

Isothermal Titration Calorimetry Measurements

ITC measurements are performed with a Microcal LLD (Northampton, MA) Omega titration calorimeter with 1.43 ml of solution and reference vessels, capable of detecting heat pulses as low as 0.8 μJ. The results are analyzed using the Origin 2.8 software program. Analysis of the measurements in terms of a binding model is straightforward as shown below, but it should be emphasized that there are some subtle corrections in the software program analysis of the ITC data in terms of these equations. For example, with each addition, there is dilution of the solution in the solution vessel so that the total protein concentration decreases monotonically with each addition of titrant and there is displacement of an equal volume of the newly mixed solution out of the fixed measurement volume of the solution vessel.

General Procedure

To increase the precision of the ITC measurements and to minimize loss of the protein mutants, the following experimental protocols are employed in the ITC measurements. For the lysozyme antigen–antibody binding measurements, the mutated antibody fragment solutions and the lysozyme solutions are dialyzed in the same buffer solution overnight at 277 K with at least two changes of buffer solution. The cNMP ligand solutions are made by dissolving the cNMP in the dialysate from the dialysis of the protein solution. Thus, the pH and salt concentrations of the ligand solution are matched to that of the corresponding protein solution. The dialysate is also used as the buffer in the reference vessel. For a series of titrations using the same buffer, the buffer solution in the reference sample is not replaced for every titration. The general measurement procedure for both the direct and indirect mutagenesis investigations is as follows.

1. A titration of just the ligand solution into the buffer solution in the solution vessel is performed for about 10 injections if there is no monotonic decrease in the observed heat of dilution. If a monotonic decrease is observed then the titration is continued for as many titrations as the planned ligand-into-protein solution titration. It is important to record the experimental results of the ligand-into-buffer titration so that it can be compared with the following ligand-into-protein solution titration.

2. The buffer solution is then removed, the solution vessel is rinsed with the dialysate several times, the protein solution is then placed into the solution vessel, and the next titration is performed with the remainder of the ligand solution in the syringe. If after four additions of the ligand solution there is little difference from the first ligand-into-buffer titration, the measurements are discontinued.

3. Before analysis of the titration results, an average value for the heat of titration is subtracted from the incremental heats for each injection. If the heat of dilution is large and exhibits a monotonic decrease with each addition of titrant than the heats of dilution for the whole run are subtracted from the titration heats of the subsequent ligand–protein titration. However, this is not usually the case, particularly if the buffers of both solutions are matched. Heats of dilution of the ligand solution are also obtained from additions of the ligand solution into the protein solution beyond the saturation level of the protein.

Error Analysis

The software programs that provide the best fit of the binding model to the experimental values in the binding isotherm yield values for K_b and $\Delta_b H°$ along with the error associated with each parameter. The error is derived from the square root of the sum of the squares of the differences between the calculated value and the experimental value. In addition to this uncertainty from the fit is an uncertainty in the concentration determined here by ultraviolet (UV) absorption measurements at 280 nm for the mutated proteins. It is estimated that this uncertainty can be as high as 3% and so the uncertainties in the values reported below are as follows:

$$\delta(\Delta_b H^\circ) = \{\delta[\Delta_b H^\circ(\text{fit})]^2 + (0.009)\Delta_b H^\circ(\text{fit})\}^{\frac{1}{2}}$$
$$\delta(K_b) \;\;\; = \{\delta[K_b(\text{fit})]^2 + (0.009)K_b(\text{fit})\}^{\frac{1}{2}}$$
$$\delta(\Delta_b G^\circ) = RT\delta(K_b)/K_b$$

and for the binding entropy

$$\delta(\Delta_b S^\circ) = \{\delta(\Delta_b H^\circ/T)^2 + \delta(\Delta_b G^\circ/T)^2\}^{\frac{1}{2}}$$

These uncertainty values are estimated upper limit uncertainties.

Analysis of Direct Mutagenesis Effects: Lysozyme–Antibody Mutant Interactions

Because antibody–antigen binding interactions are highly specific and exhibit high binding affinities, mutations at the antigen-binding site can significantly affect the thermodynamics of the binding interaction. The binding of hen egg white lysozyme (HEL) antigen to the Fv fragment of D1.3 antibody with mutations at a large hydrophobic area near the periphery of the antigen-binding site of the Fv fragment of the D1.3 antibody (FvD1.3) has been investigated with a Microcal Omega titration calorimeter.[3] FvD1.3 and mutants were expressed in *Escherichia coli* BMH 71–18 cells transformed with the pUC19-based expression vector pSW1-VHD1.3-VKD1.3 as described previously.[3] Mutagenesis of FvD1.3 consisted of replacing a large tryptophan residue ($V_L W92$) at the binding site with alanine, serine, valine, aspartate, histidine, or phenylalanine. Structures of the FvD1.3 mutants in complex with HEL were determined by X-ray crystallography at resolutions between 1.75 and 2.00 Å. Structures of the free wild-type antibody fragment and of the fragment mutant–HEL antigen complexes were resolved to high enough resolution so that actual changes in the number of water solvent molecules at the binding site could be resolved.

ITC Measurements

A typical ITC titration consists of titrating 5-μl aliquots of 0.50 mM HEL solution into about 0.025 mM FvD1.3 mutant solution at 297.15 K. The buffer solution consists of 10 mM sodium phosphate and 0.15 M NaCl at pH 7.4, and both the HEL and FvD1.3 mutants are dialyzed overnight in the same buffer. Concentrations of the solutions are determined after dialysis and before the ITC measurements by UV absorption measurements at 280 nm, using 10 g liter^{-1} extinction coefficients of 1.50 for wild-type

FvD1.3, 1.34 for the mutants, and 2.62 for HEL solutions.[3] The solutions
are not degassed before loading into the ITC because the precision of the
measurement is not improved by degassing and there is a possibility that
degassing will change the concentration of the solution. Titrations are performed with injections 3 min apart and continued beyond the saturation
level until after several injections the amount of heat exchanged is close
to the level in the HEL-into-buffer titration. A typical ITC titration of
5 μl of a 0.5 mM HEL solution into a 0.026 mM W92F FvD1.3 mutant solution is shown in Fig. 1 by the monotonic reduction in titration peak areas
on addition of the HEL solution to the W92F FvD1.3 mutant. There is little
heat exchanged after 45 min, as shown in Fig. 1, implying that beyond saturation of the fragment the buffer pH and ion concentration are the same
for both the HEL and FvD13 solutions.

ITC Analysis

In Fig. 1, the total heat exchanged per mole of ligand in the solution
vessel (Q_t) is plotted as a function of the ratio of the moles of ligand to
the number of moles of protein in the solution vessel and this resulting

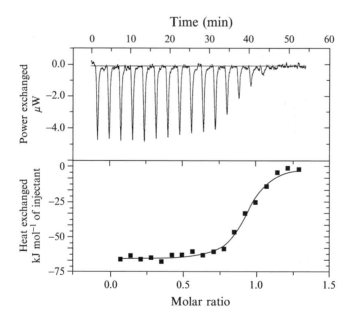

FIG. 1. *Top:* ITC titration of 5-μl aliquots of 0.5 mM HEL into a 0.026 mM concentration
of the W92→F mutant of FvD1.3 in 10 mM sodium phosphate buffer containing 0.15 M NaCl
at pH 7.4 and at 297.15 K. *Bottom:* Binding isotherm for the titration and the fit of the data to
a single-site binding model.

binding isotherm was then fitted to a simple 1:1 binding model. The conservation of mass yields the following equation for this model,[1]

$$Q_t = n[C_t]\Delta_b H^\circ V\{1 + [X_t]/n[C_t] + 1/nK_b[C_t]$$
$$- [(1 + [X_t]/n[C_t] + 1/nK_b[C_t]]^2 - 4[X_t]/n[C_t]]^{1/2}\}/2 \qquad (3)$$

where $n = 1$, the stoichiometry of the binding reaction, $[C_t]$ is the total FvD1.3 concentration in the solution vessel, $\Delta_b H^\circ$ is the binding enthalpy, V is the volume of the solution vessel, $[X_t]$ is the total HEL concentration, and K_b is the binding constant. Analysis of the fit of the binding model to the binding isotherm in Fig. 1 yielded $n = 0.91 \pm 0.01$, $K_b = (5.7 \pm 1.0) \times 10^6 \ M^{-1}$, and $\Delta_b H^\circ = -6.62 \pm 0.86$ kJ mol^{-1}. In summary, it was found that the binding affinities of HEL for the other mutants were reduced from the known value of $(5.0 \pm 0.8) \times 10^7 \ M^{-1}$ for the wild type to $(2.5 \pm 0.4) \times 10^5 \ M^{-1}$ for the valine mutant, which is close to the binding affinities for the other smaller side-chain alanine, serine, and aspartate mutants.[3] The larger side-chain tyrosine and histidine mutants exhibited binding affinities an order of magnitude higher than those of the smaller side-chain mutants. Similarly, the binding enthalpies increased from -90.0 ± 3.0 kJ mol^{-1} for wild type to -60.7 ± 2.1 kJ mol^{-1} for the serine mutant, but no trend was observed between the size of the side chain and the increase in binding enthalpy, in contrast to that observed for the binding affinities. For example, the smaller aspartate side chain exhibited a lower binding enthalpy (-74.1 ± 3.8 kJ mol^{-1}) than that of the tyrosine side chain (-66.1 ± 2.0 kJ mol^{-1}). The binding entropy was determined from Eq. (1) and it was observed that a plot of $\Delta_b H^\circ$ versus $T\Delta_b S^\circ$ was linear, that is, loss in binding enthalpy is partly compensated by a gain in binding entropy. This observation of enthalpy–entropy compensation implies that water is involved in the binding mechanism as indeed was observed by the presence of water molecules in the binding interface between the mutant and the HEL.

Relationship of Thermodynamic Changes to Structural Changes

From X-ray crystal structures of the mutants, solvent-accessible surface areas were calculated for residues $V_L 91$–$V_L 94$, which constitute a loop of solvent-exposed residues in the unbound state and include the mutation at $V_L 92$. This was done instead of calculating the change in solvent-accessible surface area for the entire binding site interface because the integrity of the binding site structure was basically maintained and isomorphism between the wild-type and mutant complex structures introduced additional, unnecessary errors in the calculation of the areas.[3] Changes in the solvent-accessible surface areas on binding were calculated by the programs

AREAIMOL and DIFFAREA from the CCP4 suite of programs,[4] using a probe radius of 1.4 Å.

Changes in the thermodynamic binding quantities were then analyzed in terms of changes in the solvent-accessible surface areas of the nonpolar amino acid residues ($\Delta\Delta ASA_{nonpolar}$) and the polar amino acid residues ($\Delta\Delta ASA_{polar}$) on binding of HEL to the Fv fragment mutant. This was based on the observation, for example, that for a hydrophobic interaction changes in the binding free energy are expected to correlate linearly with $\Delta\Delta ASA_{nonpolar}$.[5] It was observed, however, that changes in the binding free energy changes, $\delta(\Delta_b G^\circ) = \Delta_b G^\circ$ (mutant) $-\Delta_b G^\circ$ (wild type), correlated linearly only with changes in the solvent-accessible surface areas of the nonpolar amino acid residues as shown in Fig. 2, where $\delta(\Delta_b G^\circ)$ is plotted as a function of $\Delta\Delta ASA_{nonpolar}$. For comparison, a similar plot of $\delta(\Delta_b G^\circ)$ as a function of $\Delta\Delta ASA_{polar}$ is also shown in Fig. 2. It is apparent that there is no correlation between $\delta(\Delta_b G^\circ)$ and ΔASA_{polar} in Fig. 2. For the contribution from direct mutagenesis of $V_L 92$, it was found that

$$\delta(\Delta_b G^\circ) = 88\Delta\Delta ASA_{nonpolar}(\text{J mol}^{-1}\text{Å}^{-2}) \tag{4}$$

as would be expected for an exclusively hydrophobic interactions at $V_L 92$. Although values for the binding heat capacity change could be determined from ITC measurements, in this investigation they were calculated to conserve material. Values for $\Delta_b C_p$ were calculated from nonpolar and polar amino acid residue contributions, using the following equation[2]:

$$\Delta_b C_p = \Delta_b C_{p, nonpolar} + \Delta_b C_{p, polar} \tag{5a}$$

and values for $\Delta_b C_{p, nonpolar}$ and $\Delta_b C_{p, polar}$ were calculated from changes in the solvent-accessible surface areas (in angstroms squared) and T (in degrees Kelvin) as shown:

$$\Delta_b C_{p, nonpolar} (\text{J mol}^{-1} \text{ K}^{-1}) = [1.88 + (11.0 \times 10^{-4})(T - 298.15)$$
$$- (17.6 \times 10^{-5})(T - 298.15)^2] \tag{5b}$$

$$\Delta\Delta ASA_{nonpolar}$$

$$\Delta_b C_{p, nonpolar} (\text{J mol}^{-1} \text{ K}^{-1}) = [-1.09 + (11.9 \times 10^{-4})(T - 298.15)$$
$$- (18.0 \times 10^{-5})(T - 298.15)^2] \tag{5c}$$

$$\Delta\Delta ASA_{polar}$$

[4] Collaborative Computational Project. *Acta Crystallogr. D Biol. Crystallogr.* **50,** 760 (1994).
[5] C. H. Chothia, *Nature* **248,** 338 (1974).

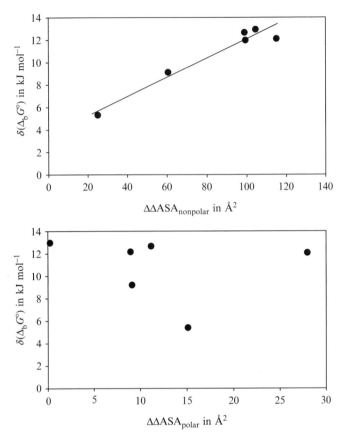

Fig. 2. A plot of $\Delta_b G°$ (mutant) $-$ $\Delta_b G°$ (wild type) versus differences in solvent-accessible surface areas of nonpolar residues ($\Delta\Delta ASA_{nonpolar}$) and solvent-accessible surface areas of polar residues ($\Delta\Delta ASA_{polar}$), between mutant and wild-type FvD13 fragments at 297.15 K.

The solvent contribution to the binding entropy, $\Delta_b S_{solv}$, is related to the calculated components of the heat capacity change for the binding reaction ($\Delta_b C_p$) through[2]

$$\Delta_b S_{solv} = \Delta_b C_{p,\,nonpolar} \ln\left(T/385.15\right) + \Delta_b C_{p,\,polar} \ln\left(T/335.15\right) \quad (6)$$

However, differences in the binding entropy change between the mutants, $\delta(\Delta_b S°)$, linearly correlated well only with changes in the solvent-accessible nonpolar surface area in Eq. (6) at 297.15 K. Because these losses also correlated well with a decrease in the binding free energy, $\Delta_b G°$, as shown in Fig. 2, it was then concluded that the exclusion of water from the binding interface was a predominant driving force of the binding mechanism, as would be expected for a hydrophobic interaction.[3]

Comments

This example illustrates the success of analyzing changes in binding thermodynamics with changes in the solvent-accessible surface area of polar and nonpolar amino acid residues at the binding site determined from detailed structural information on the molecular level. High-resolution X-ray crystal structures were necessary for accurate determination of changes in the solvent-accessible surface area. Other thermodynamic functions can be analyzed by the same method. For example, the binding enthalpy has also been related to changes in the solvent-accessible surface area through the following equations[2]:

$$\Delta_b H(T = 333 \text{ K})(\text{J mol}^{-1}) = -35.52 \Delta \Delta \text{ASA}_{\text{nonpolar}} + 131.4 \Delta \Delta \text{ASA}_{\text{polar}}$$

$$(7)$$

where $\Delta \Delta \text{ASA}$ is in Å^{-2} and with

$$\Delta_b H(T) = \Delta_b H (T = 333 \text{ K}) + \Delta C_p(T - 333) \qquad (8)$$

the binding enthalpy can be determined at any temperature. It should be emphasized that X-ray crystallographic data of high resolution (≥ 1.9 Å) is necessary for accurate calculation of $\Delta \Delta \text{ASA}$. It is also important to minimize the number of amino acid residues involved in the surface area calculation because too large an area calculation may introduce additional and unnecessary uncertainty in the calculation. A probe radius of 1.4 Å is usually used for the water molecule in the CCP4 suite of programs; a different probe radius would change $\Delta \Delta \text{ASA}$. Furthermore, the thermodynamic quantities need to be determined with a high-enough level of precision to observe δ values that are larger than the experimental uncertainties. After values of $\Delta \Delta \text{ASA}$ for the polar and nonpolar residues are calculated for a series of mutations at the binding site, then changes in the thermodynamic quantities, $\delta(\Delta_b G^\circ)$, $\delta(\Delta_b H^\circ)$, and $\delta(\Delta_b S^\circ)$, can be plotted as a function of $\Delta \Delta \text{ASA}_{\text{polar}}$ and $\Delta \Delta \text{ASA}_{\text{polar}}$ to determine whether there is a linear correlation between changes in the thermodynamic quantities and changes in the solvent-accessible surface area. Then Eqs. (5) to (8) can be employed for those δ values that exhibit a linear correlation with changes in the solvent-accessible surface area to elucidate the nature of the interaction on the molecular level. These equations are based on empirical determinations that are being continually improved; the more recent literature should be continually consulted when using these relationships. Interestingly, these

equations were originally developed to relate differences in the thermodynamic quantities that describe the thermal unfolding of proteins in terms
of differences in changes of the solvent-accessible surface area of the protein
as it unfolds.

Indirect Mutagenesis: Effect of T127→L Mutation on Cooperative Binding of cAMP to CRP

Allosteric activation of 3',5'-cyclic adenosine monophosphate (cAMP)
receptor protein (CRP) by cAMP results in the enhancement of transcription of more than 25 operons encoding enzymes involved in catabolite
metabolism. CRP consists of two identical subunits, each 22,500 gmol^{-1},
with an α-helical motif along the monomer–monomer interface between
the amino- and carboxyl terminal domains of each subunit.[6] The activator
cAMP binds in the amino-terminal domains of CRP and activates a conformational change in CRP so that its carboxyl-terminal domains bind specifically to a site in the promoter region of the operon, adjacent to the RNA
polymerase-binding site. This enhancement has been attributed to (1) an
increase in the binding affinity of RNA polymerase for the promoter by
CRP,[7,8] (2) bending of the promoter by CRP, resulting in more contacts between the RNA polymerase and the promoter,[9] and/or (3) an increase in
the isomerization rate of the RNA polymerase–promoter complex to a
transcriptionally active form by CRP.[8] In the cAMP-ligated state, the two
subunits of CRP in the X-ray crystal structure exhibit slightly different conformations, with the carboxyl-terminal domain bent away from the subunit
interface in one subunit (an "open" form) and the other subunit with its
carboxyl-terminal domain bent toward the interface (a "closed" form).[6]
The initiation of *in vitro* transcription of the operons is then enhanced by
more than an order of magnitude by cAMP-ligated CRP, similarly to that
observed in *in vivo* transcription.[10] *In vitro* and *in vivo* transcription activation assays show that mutations along the subunit interface of CRP
significantly alter the enhancement of the activation of transcription by
CRP.[10,11] For example, a mutation of T127 to L, which converts the
α-helical interface to a more perfect leucine zipper motif, results in the

[6] I. T. Weber and T. A. Steitz, *J. Mol. Biol.* **198**, 311 (1987).
[7] S. Busby and R. H. Ebright, *J. Mol. Biol.* **293**, 199 (1999).
[8] D. C. Straney, S. B. Straney, and D. M. Crothers, *J. Mol. Biol.* **206**, 41 (1989).
[9] S. Dethiollaz, P. Eichenberger, and J. Geiselmann, *EMBO J.* **15**, 5449 (1996).
[10] S. Wang, Y. Shi, I. Gorshkova, and F. P. Schwarz, *J. Biol. Chem.* **275**, 33457 (2000).
[11] S. F. Leu, C. H. Baker, E. J. Lee, and J. G. Harman, *Biochemistry* **38**, 6222 (1999).

enhancement of transcription in the absence of cNMP and by an analog of cAMP, 3',5'-cyclic guanosine monophosphate (cGMP).[10]

ITC Measurements

CRP and mutated CRP protein, 127T→L (T127L), solutions are employed at concentrations in the range of 0.1 to 0.5 mM in 0.05 M potassium phosphate–potassium hydroxide buffer containing 0.2 mM dithiothreitol (DDT), 0.2 mM sodium EDTA, 5% (by volume) glycerol, and from 0.2 to 0.5 M KCl and dialyzed in the buffer overnight.[12,13] The concentrations of the CRP and of the mutant are determined after dialysis and before the titration by UV absorption measurements at 280 nm, using an extinction coefficient of 3.5×10^4 cm M^{-1}. The cNMP solutions are made up at concentrations 20-fold the number of binding sites on the protein concentration (2 per mole of CRP) by adding appropriate amounts of the sodium cNMP salt to the final dialysate and the concentrations are determined by UV absorption measurements, with extinction coefficients of 1.23×10^4 cm M^{-1} at 260 nm for cAMP and 1.34×10^4 cm M^{-1} at 250 nm for cGMP.[12] The ITC titrations are performed with a Microcal Omega ITC isothermally at 294.15 and 312.15 K with 5-μl injections of the cNMP solution that are 3 min apart. The binding isotherm for a titration of 8.0 mM cAMP solution into 0.15 mM wild-type CRP in the potassium phosphate buffer is shown in Fig. 3 (top) and appears to be more complex than the simple 1:1 binding isotherm of the titration of HEL to the FvD1.3 mutant shown in Fig. 1. The titration is discontinued after 70 min because the areas of the titration peaks after 60 min are the same as with the previous titration of the cAMP solution into the buffer solution. For titrations of the 8.0 mM cAMP solution into the buffer and into the CRP, there is a heat of dilution that must be subtracted from the titration heats observed for titration of the cAMP solution into the CRP solution before analysis.

ITC Analysis

In Fig. 3 (top), an increase to a maximum value is first observed in the integrated endothermic peak areas and the maximum value is then followed by a monotonic decrease in the peak areas to the baseline as the titration is continued. The unusual binding isotherm was fitted to a two-site

---

[12] I. Gorshkova, J. L. Moore, K. H. McKenney, and F. P. Schwarz, *J. Biol. Chem.* **270,** 21679 (1995).

[13] J. L. Moore, I. Gorshkova, J. W. Brown, K. H. McKenney, and F. P. Schwarz, *J. Biol. Chem.* **271,** 21273 (1996).

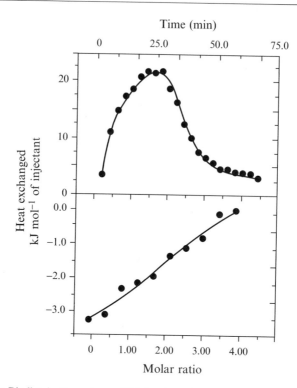

Fig. 3. *Top:* Binding isotherm for an ITC titration of 5-μl aliquots of 8.0 mM cAMP into 0.15 mM CRP in 0.5 M KCl–50 mM potassium phosphate buffer with EDTA and DDT (0.2 mM each) and 5% glycerol at pH 7.0 at 297.15 K. The fit of the data to a two-site interacting binding model is shown by the solid line. *Bottom:* Binding isotherm for an ITC titration of 5-μl aliquots of 6.7 mM cAMP into a 0.28 mM concentration of the T127→L mutant of CRP in 0.5 M KCl–50 mM potassium phosphate buffer with EDTA and DDT (0.2 mM each) and 5% glycerol at pH 7.0 at 296.15 K. The titrant injections were 6 min apart for this titration. The fit of the data to a single site binding model is shown by the solid line.

interactive binding model, which yields a binding constant and a binding enthalpy for each of the two subunit sites. For an interacting two-site mechanism,

$$Q_t = C_t V \{K_b(1)[X_t]\Delta_b H°(1) + K_b(1)K_b(2)[X_t]^2\{\Delta_b H°(1) + \Delta_b H°(2)\}\}/P$$

$$(9a)$$

where

$$P = 1 + K_b(1)[X_t] + K_b(1)K_b(2)[X_t]^2 \qquad (9b)$$

In this equation, $K_b(1)$ and $\Delta_b H°(1)$ are the binding constant and enthalpy change, respectively, for binding of cAMP to the first site, and $K_b(2)$ and $\Delta_b H°(2)$ are the binding constant and enthalpy change, respectively, for binding to the second site. The fitting programs usually yield the on-site binding constants that are different from the macroscopic binding constants. For a two-site binding mechanism, the macroscopic binding constants to the first site and to the second site are then, respectively, $2K_b(1)$ and $K_b(2)/2$. A coefficient of cooperativity, α, then can be defined for any two-site model as the ratio of $K_b(2)$ to $K_b(1)$ and the energy of interaction can be defined as $\Delta G = -RT \ln\{\alpha\}$. For an identical site mechanism in the absence of cooperativity, $K_b(1) = K_b(2)$ and $\alpha = 1$. Because the fitting procedure of the model to the binding isotherm involved four parameters, the fitting procedure was facilitated by choosing initial values for the binding constants close to those found in the literature employing equilibrium dialysis.[14] To improve the fits for some of the titrations, it was necessary to reduce the CRP concentration by about 10%, which was the lower error limit in the CRP protein assays. In summary, the fitting results were best described in terms of an exothermic binding interaction to the first site with $K_b(1) = (2.9 \pm 0.4) \times 10^4 \ M^{-1}$ and $\Delta_b H°(1) = -5.6 \pm 0.6$ kJ mol^{-1} at 297.15 K and 0.5 M KCl followed by an endothermic binding reaction to the second site with $K_b(2) = (14.0 \pm 1.0) \times 10^4 \ M^{-1}$ and $\Delta_b H°(1) = 48.1 \pm 2.1$ kJmol^{-1} at 0.5 M KCl.[12] The relative values of the on-site binding constants showed that the binding interaction exhibited a positive cooperativity coefficient that decreased from $\alpha = 11.7 \pm 2.3$ to 4.8 ± 0.7 with an increase in KCl concentration from 0.2 to 0.5 M. In contrast to this complex binding isotherm, the binding of cAMP to the T127L mutant of CRP was observed to follow a simple site-independent binding mechanism.[12,13] The binding isotherm for this titration is shown in Fig. 3 (bottom) by the simple monotonic reduction in the integrated titration peak areas on addition of 5-μl aliquots of a 6.7 mM cAMP solution, 6 min apart, to a 0.28 mM T127L CRP mutant solution at 296.15 K. (The rate of decrease in the integrated peak areas per addition of titrant is more gradual in Fig. 3 than in Fig. 1 because, although the cAMP concentration is 20 times greater than the HEL concentration, the HEL binding constant is two orders of magnitude higher than that of cAMP binding to the T127L CRP mutant.) The fit of the independent two site-binding model to the experimental binding isotherm is also shown in Fig. 3. The binding reaction is exothermic with $K_b = (8.0 \pm 0.8) \times 10^4 \ M^{-1}$ and $\Delta_b H° = -3.3 \pm 0.3$ kJ mol^{-1} at 296.15 K and 0.5 M KCl, the stoichiometry $n = 2.0 \pm 0.1$, so that

---

[14] M. Takahashi, B. Blazy, and A. Baudras, *Biochemistry* **19**, 5124 (1980).

one cAMP binds to each subunit of the two subunits of CRP, and the positive cooperativity between the binding sites for cAMP binding to CRP is no longer evident. Furthermore, the analog cGMP binding measurements to CRP and to the T127L mutant were also observed to be site independent and exothermic, for example, $K_b = (1.75 \pm 0.18) \times 10^4 \, M^{-1}$ and $\Delta_b H° = -7.1 \pm 0.3 \, kJ \, mol^{-1}$ for cGMP binding to CRP at 293.15 K and 0.5 M KCl.[12]

ITC measurements were also performed at different temperatures for determination of the heat capacity change for each binding site and at different pH levels to further investigate differences between the CRP and T127L mutant binding mechanisms. It was found that the heat capacity change for binding to the second site in CRP was $-1.47 \pm 0.17 \, kJ \, mol^{-1} \, K^{-1}$, about 5-fold larger than that observed for binding to the first site $(-0.30 \pm 0.08 \, kJ \, mol^{-1} \, K^{-1})$.[12] In summary, for cNMP binding to the first site in CRP and to both sites in the T127L mutant, the binding is exothermic and proceeds with a binding affinity of about $10^4 \, M^{-1}$. However, endothermic binding with a large heat capacity change occurs only to the second site of CRP. When measurements of cAMP binding to the T127L mutant were performed at pH 5.2 instead of pH 7.0, the two site-dependent binding isotherm observed for CRP at neutral pH reappeared for the T127L mutant along with a larger heat capacity change of $-0.95 \pm 0.10 \, kJ \, mol^{-1} \, K^{-1}$ for binding to the first site and a positive heat capacity change of 0.42 $\pm 0.10 \, kJ \, mol^{-1} \, K^{-1}$ for binding to the second site of the T127L mutant.[15] This implies that cAMP binding to the second site of wild-type CRP and to the T127L mutant at pH 5.2 results in a conformational change and this conformational change is substantially reduced or eliminated at neutral pH by the T127L mutation at the CRP subunit interface.

Relationship between Thermodynamic Changes and Indirect Mutagenesis

The effects of mutagenesis at the subunit interface of CRP on the thermodynamics of cNMP binding to CRP were analyzed in terms of subtle changes in the structure of CRP. Unfortunately, the structure of free CRP has not been determined by X-ray crystallography, whereas structural resolutions as high as 1.8 Å can be obtained with cAMP-ligated CRP crystals. SANS was then employed to obtain structural information at much lower resolution by comparing the neutron scattering of free CRP, CRP saturated with cAMP, and a free and saturated cAMP double mutant of CRP that contains an additional mutation along the subunit interface of

[15] Y. Shi, S. Wang, and F. P. Schwarz, *Biochemistry* **39**, 7300 (2000).

S128→A in addition to the T127→L mutation.[16] (The double mutant was used in this investigation because there was too much scatter in the neutron-scattering data with the T127L mutant solutions at the high concentrations necessary for SANS measurements in D_2O solvent. There was some aggregation of the T127L mutant in the D_2O buffer at these high concentrations.) The scattering results were then compared with simulated SANS scattering curves generated from the energy-minimized X-ray crystal structure of the cAMP-ligated CRP complex[6] with the cAMP removed and with both subunits in the open form as well as with both subunits in the closed form. The solutions ranged from 0.04 to 0.14 mM in protein concentration and the cAMP-saturated solutions contained from 0.2 to 17 mM cAMP. SANS results are usually plotted in terms of the neutron-scattering intensity versus the sine of the neutron-scattering angle. The neutron-scattering intensity at the angle of incidence yields the radius of gyration of the solute. SANS data plots of unligated CRP in solution showed good agreement with the calculated results for CRP with both subunits in the open form. SANS data plots for the cAMP-saturated CRP solutions exhibited good agreement with the calculated results for both subunits in the closed form. This is further confirmed by the radius of gyration: the experimental value for unligated CRP is 22.0 ± 0.2 Å, which is in better agreement with the calculated value of 20.5 ± 0.4 Å for the exclusively open form than with the calculated value of 18.7 ± 0.3 Å for the exclusively closed form of CRP.[16] The SANS data plots for the free double mutant were the same as for the cAMP-saturated double mutant, which implies that there is little structural change between the free double mutant and the cAMP-ligated double mutant.[16] This lack of significant structural change for the double mutant on cAMP binding agrees with the thermodynamic cAMP-binding results for the double mutant and the T127L mutant, which do not exhibit any evidence of the large conformational change observed for cAMP binding to the second site of CRP. An X-ray crystallography structure of the cAMP-ligated CRP bound to a short DNA duplex that contained the 26 bp-binding site sequence of the promoter showed that both subunits of CRP are in the closed form.[17] An X-ray crystallography structure of the cAMP-ligated double mutant[18] exhibits a conformation that has both the open subunit and closed subunit as found in the cAMP-ligated wild-type CRP structure. However, the open

[16] S. Krueger, I. Gorshkova, J. Brown, J. Hoskins, K. H. McKenney, and F. P. Schwarz, *J. Biol. Chem.* **273**, 20001 (1998).

[17] S. C. Schultz, G. C. Shields, and T. A. Steitz, *Science* **253**, 1001 (1991).

[18] S. Chu, M. Tordova, G. L. Gilliland, I. Gorshkova, Y. Shi, S. Wang, and F. P. Schwarz, *J. Biol. Chem.* **276**, 11230 (2001).

subunit structure of the double mutant is more aligned with the structure of the closed subunit in wild-type cAMP-ligated CRP. That the double-mutant structure is closer to the activated form is evident in the observation that the double mutant can enhance the activation of transcription in the absence of cNMP almost to the same level as cAMP-ligated CRP.[10] The structural information implies that the conformational change from free wild-type CRP to the active DNA-bound state involves both subunits undergoing a change from a conformation in which both subunits are in the open form to the exclusively closed form in solution on saturation with cAMP in solution.

To further validate the nature of the structural change induced by the T127→L mutation of CRP, circular dichroism (CD) was employed to investigate in particular the nature of the structural change of the T127L mutant as a function of pH.[15] CD spectrum measurements were performed on the T127L mutant and, for comparison, on CRP, using a Jasco (Easton, MD) J-720 spectropolarimeter operated with 0.1-cm cells for measurements between 350 and 250 nm. The CRP and T127L mutant solutions were 0.125 mM and the CD spectra were recorded at 297.15 K. The CD spectra above 250 nm exhibit different changes in tertiary structure between CRP and the T127L mutant as the pH is lowered from pH 7.0 to 5.2, whereas the absorption peak around 295 nm remains the same.[15] The constancy of the peak at 295 nm was interpreted to mean that the tertiary structure around the two tryptophan residues in the amino-terminal domain remains the same with change in pH. This implies that the pH change has little effect on the structure of the amino-terminal domains, and therefore that any implied structural change in the tertiary structure would occur in the C-terminal domains, and that these changes with change in pH are different for CRP and the T127L mutant.[15] Again, this structural change in the C-terminal domains of T127L may result from the mutant changing from an almost exclusively closed form at pH 7.0 to one with both subunits almost in the open form at pH 5.2, as observed for unligated CRP at pH 7.0 by the SANS results. This change toward the open form for unligated T127L at pH 5.2 would account for the cooperativity of cAMP binding to T127L at this pH.

Comments

Analysis of the effects of indirect T127→L mutagenesis on the thermodynamic changes in cooperativity of cAMP binding to CRP has relied heavily on analysis of the structural changes induced in CRP by this mutation. This analysis has resulted in the development of a structural model that best describes how the conformational changes induced by the

T127→L mutation result in the observed thermodynamic changes in the cNMP–CRP binding interaction. The structural methods employed in this analysis of indirect mutagenesis effects included X-ray crystallography, SANS, and CD measurements. Furthermore, it should be emphasized that this structural model offers an explanation based on the presently available structural and thermodynamic data about this system and could, thus, be changed as more experimental data about this intriguing system become available. Analysis of the effect of indirect mutagenesis on the thermodynamics of protein–ligand interactions involves development of a structural model to explain how mutations away from the binding site change the observed thermodynamics of the binding reaction: this is not as straightforward as analysis of the effect of direct mutagenesis on the thermodynamics of protein–ligand interactions in terms of changes in the solvent-accessible surface areas at the binding site.

Acknowledgments

Certain commercial materials, instruments, and equipment are identified in this chapter in order to specify the experimental procedure as completely as possible. In no case does such identification imply a recommendation or endorsement by the National Institute of Standards and Technology nor does it imply that the materials, instruments, or equipment identified is necessarily the best available for the purpose.

[8] Multiple Binding of Ligands to a Linear Biopolymer

By Yi-der Chen

Introduction

Many biological processes involve cooperative binding of several different protein ligands to linear biopolymers (such as DNA, actin, and microtubules). If each subunit (monomer) of the biopolymer has only one ligand-binding site, the binding is purely competitive and cooperative binding phenomena are derived mainly from nearest-neighbor ligand–ligand interactions. This kind of binding is referred to as the "single-binding" type. On the other hand, in "multiple-binding" systems in which each subunit of the biopolymer contains two or more binding sites, so that more than one ligand can bind to the same subunit simultaneously, the cooperative binding of one ligand may be modulated allosterically by the other ligand. For example, binding of myosin subfragment 1 (S-1) molecules to a bare actin is known to be noncooperative. But, when the actin is prebound with tropomyosin–troponin molecules, the binding of S-1 becomes coopera-tive.[1-4] Theoretical treatment of cooperative binding of ligands to linear biopolymers is an interesting biophysical problem and has attracted consid-erable attention.[5-12] However, most of the formalisms developed so far are concentrated on the "single-binding" model. This chapter introduces a formalism that can be used to treat not only single-binding systems, but also multiple-binding systems.[13-15] Instead of discussing the theoretical basis of the formalism, this chapter uses a simple two-ligand multiple-binding system to illustrate how to carry out the evaluation of the secular equation and the binding isotherms. The purpose is to present the general methodology of the formalism so that one may be able to modify or extend when needed.

Model and Parameters

The two-ligand multiple-binding model shown in Fig. 1A is used to pre-sent the formalism. Ligand I can cover one monomer whereas ligand II can cover three monomers when bound to the biopolymer. Ligand II has a head part (represented by a vertical line as shown in Fig. 1A) and a tail part and is assumed to bind to the biopolymer with a fixed direction (the head always points toward the right). The effect of ligand II on the binding of ligand I is not uniform among the three subunits covered by a ligand II: ligand I can bind to the two subunits covered by the tail part of a ligand II, but not to the subunit covered by the head part (the "mosaic" model). As shown in Fig. 1A, there are a total of four cooperativity param-eters in this model: y_1 (I–I), y_2 (II–II), y_{12} (I–II), and y_{21} (II–I). One must note that y_{12} may be different from y_{21} because the head and the tail of ligand II may interact differently with ligand I. The other parameters that

[1] L. E. Greene and E. Eisenberg, *Proc. Natl. Acad. Sci. USA* **77,** 2616 (1980).
[2] T. L. Hill, E. Eisenberg, and L. E. Greene, *Proc. Natl. Acad. Sci. USA* **77,** 3186 (1980).
[3] Y. Chen, B. Yan, J. M. Chalovich, and B. Brenner, *Biophys. J.* **80,** 2338 (2001).
[4] A. M. Resetar, J. M. Stephens, and J. M. Chalovich, *Biophys. J.* **83,** 1039 (2002).
[5] D. Poland, "Cooperative Equilibrium in Physical Biochemistry." Clarendon, Oxford, 1979.
[6] T. L. Hill, "Cooperativity Theory in Biochemistry." Springer-Verlag, New York, 1985.
[7] A. Ben-Naim, "Cooperativity and Regulation in Biochemical Processes." Kluwer-Academic/Plenum, 2001.
[8] S. Lifson, *J. Chem. Phys.* **40,** 3705 (1964).
[9] Y. Chen, A. Maxwell, and H. V. Westerhoff, *J. Mol. Biol.* **190,** 201 (1986).
[10] Y. D. Nechipurenko and G. V. Gursky, *Biophys. Chem.* **24,** 195 (1986).
[11] W. Bujalowski, T. M. Lohman, and C. F. Anderson, *Biopolymers* **28,** 1637 (1989).
[12] Y. Chen, *Biophys. Chem.* **27,** 59 (1986).
[13] Y. Chen, *Biopolymers* **30,** 1113 (1990).
[14] Y. Chen and J. M. Chalovich, *Biophys. J.* **63,** 1063 (1992).
[15] B. Yan, A. Sen, J. M. Chalovich, and Y. Chen, *Biochemistry* **42,** 4208 (2003).

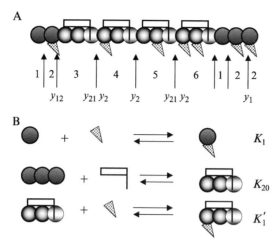

Fig. 1. The two-ligand multiple-binding model. Ligand I binds to the down side of the biopolymer and can cover only one subunit when bound, whereas ligand II binds to the up side of the biopolymer and can cover three subunits. The subunit bound by the head of ligand II, as indicated by each vertical line, loses the ability to bind ligand I. (A) A biopolymer bound with ligands showing all possible ligand-binding configurations. Each unit between two arrows is defined as an "elementary unit" of the binding system. Cooperativity parameters between two elementary units are shown under the arrows. The index of the elementary unit is shown between arrows. (B) Binding reactions that are involved in the formalism. The K values are the equilibrium binding constants.

the formalism needs are the binding constants of the three binding reactions shown in Fig. 1B. One must note that values of K_1, K_{20}, y_1, and y_2 can be evaluated using the binding isotherm measured separately for each individual ligand. Thus, the only unknown parameters in the formalism are y_{12}, y_{21}, and K_1'. One must also note that K_1' controls the allosteric effect of ligand II on the binding of ligand I and exists only in multiple-binding systems. In purely competitive, single-binding systems, cooperative binding can be modulated only by the cooperativity parameters y_{12} and y_{21}.

Let us define δ as

$$\delta \equiv K_1'/K_1 \tag{1}$$

Then, the value of δ determines the strength of the allosteric effect between the two ligands. That is, when $\delta = 1$, ligand II has no allosteric effect on the binding of ligand I and vice versa. On the other hand, the binding of one ligand will enhance or reduce the binding of the other depending on whether δ is larger or smaller than 1. At the extreme limit of $\delta = 0$, subunits covered by ligand II can no longer bind ligand I and vice versa. As a result, this multiple-binding model behaves like a single-binding, purely competitive system, although the two ligands still bind to different binding sites.

The Formalism

Step 1: Identify Elementary Units

An elementary unit of a binding system is a complex between the biopolymer and the ligands that does not contain any free bond (the bond between two neighboring subunits that is not covered by a ligand, as indicated by arrows in Fig. 1A). Thus, the simplest elementary unit of the system is the single bare subunit with no bound ligand. This unit is indexed as unit 1 of the set and is the only one in the set that has no bound ligand. The rest of the elementary units contain either bound ligand I or bound ligand II, or both. As shown in Fig. 1A, the binding system contains six elementary units although it has only two ligand species. In single-binding systems, the total number of elementary units is exactly equal to the number of ligand species in the system.

Step 2: Evaluate Statistical Weight of Isolated Elementary Unit

The weight of each elementary unit is assigned using the following rules: (1) the weight of elementary unit 1 (the bare subunit) is assigned with a weight of "unity" ($x_1 = 1$); (2) the weight of units containing only one bound ligand is equal to the product of the binding constant and the concentration of the ligand (the binding potential[13]). For example, as shown in Fig. 2, we have $x_2 = X_I(\equiv K_1 C_I)$ and $x_3 = X_{II}(\equiv K_{20} C_{II})$; and (3) the weight of a unit containing one ligand II and n ligand I with m pairs of nearest neighbors is equal to $(\delta X_I)^n X_{II}(y_1)^m$. For example, unit 6 has two ligand I molecules and one pair of nearest neighbors. As a result, we have $x_6 = (\delta X_I)^2 X_{II} y_1$. Two points we like to emphasize here. First, rule 3 is needed only for elementary units that contain more than one bound ligand. Therefore, this rule is not needed for single-binding systems. Second, it is important to include the cooperativity parameter in the weight of an elementary unit if ligand–ligand interactions exist in that unit (such as in unit 6). Without the inclusion of the cooperativity parameter in the weight, the matrix formalism presented below will not work properly.

Step 3: Construct Matrix A

With the set of elementary units obtained and their statistical weights assigned according to steps 1 and 2 above, the binding isotherms of this multiple-binding system can be evaluated by a matrix method similar to

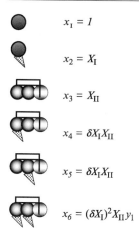

$$x_1 = 1$$

$$x_2 = X_I$$

$$x_3 = X_{II}$$

$$x_4 = \delta X_I X_{II}$$

$$x_5 = \delta X_I X_{II}$$

$$x_6 = (\delta X_I)^2 X_{II} y_1$$

FIG. 2. The statistical weight of an isolated elementary unit used in the formalism. X_I ($\equiv K_1 C_I$) and X_{II} ($\equiv K_{20} C_{II}$) are the "binding potential" of ligand I and ligand II, respectively. The value of δ ($\equiv K_1'/K_1$) determines the effect of bound ligand II on the binding of ligand I. K_1' and K_1 are, respectively, the binding constant of ligand I to a subunit in the presence and absence of a bound ligand II, and K_{20} is the binding constant of ligand II to a set of three bare biopolymer subunits. C_I and C_{II} are, respectively, the concentrations of ligand I and ligand II in solution.

that used in treating classic linear Ising lattice problems.[5,6] That is, we first construct matrix **A** that specifies the "exact" statistical weight of an elementary unit in a pair formation and then evaluate the "secular" equation from this **A** matrix. Construction of the **A** matrix contains the following two operations: (1) place every element on the ith row of **A** with the weight of the elementary unit i ($i = 1, 6$) obtained in step 2 above; and then (2) multiply each **A** (i, j) with the cooperativity factor between unit i and unit j. The matrix **A** thus constructed for the model in Fig. 1 is shown in Fig. 3.

Step 4: Secular Equation

The way to obtain the secular equation from the **A** matrix in Fig. 3 for this binding system is slightly different from that used in Ising systems, because ligand II in this system can cover more than one biopolymer subunit (n-mer). Thus, instead of subtracting γ (a variable) from every diagonal element of **A** as when every unit is a monomer, we subtract γ^{n_i} from **A** (i, i), where n_i represents the number of biopolymer subunits contained in elementary unit i (e.g., in this system we have $n_1 = n_2 = 1$ and $n_3 = n_4 =$

$$
\mathbf{A} =
\begin{bmatrix}
x_1 & x_1 & x_1 & x_1 & x_1 & x_1 \\
x_2 & x_2 y_1 & x_2 y_{12} & x_2 y_1 y_{12} & x_2 y_{12} & x_2 y_1 y_{12} \\
x_3 & x_3 y_{21} & x_3 y_2 & x_3 y_2 y_{21} & x_3 y_2 & x_3 y_2 y_{21} \\
x_4 & x_4 y_{21} & x_4 y_2 & x_4 y_2 y_{21} & x_4 y_2 & x_4 y_2 y_{21} \\
x_5 & x_5 y_{21} & x_5 y_2 & x_5 y_2 y_{21} & x_5 y_2 & x_5 y_2 y_{21} \\
x_6 & x_6 y_{21} & x_6 y_2 & x_6 y_2 y_{21} & x_6 y_2 & x_6 y_2 y_{21}
\end{bmatrix}
$$

FIG. 3. The matrix $\mathbf{A}$ of the model. Each $\mathbf{A}$ (i, j) is the product of the weight of elementary unit i and the cooperativity parameter between unit i and unit j when unit j is at the right side of unit i.

$n_5 = n_6 = 3$ as shown in Fig. 1A). The secular equation of this binding system is easily obtained as

$$
\begin{vmatrix}
x_1 - \gamma & x_1 & x_1 & x_1 & x_1 & x_1 \\
x_2 & x_2 y_1 - \gamma & x_2 y_{12} & x_2 y_1 y_{12} & x_2 y_{12} & x_2 y_1 y_{12} \\
x_3 & x_3 y_{21} & x_3 y_2 - \gamma^3 & x_3 y_2 y_{21} & x_3 y_2 & x_3 y_2 y_{21} \\
x_4 & x_4 y_{21} & x_4 y_2 & x_4 y_2 y_{21} - \gamma^3 & x_4 y_2 & x_4 y_2 y_{21} \\
x_5 & x_5 y_{21} & x_5 y_2 & x_5 y_2 y_{21} & x_5 y_2 - \gamma^3 & x_5 y_2 y_{21} \\
x_6 & x_6 y_{21} & x_6 y_2 & x_6 y_2 y_{21} & x_6 y_2 & x_6 y_2 y_{21} - \gamma^3
\end{vmatrix} = 0
\tag{2}
$$

where $|\bullet|$ denotes a determinant.

After carrying out the following two sets of operations: (1) subtracting column 6 from column 4 and column 5 from column 3, followed by (2) adding row 6 to row 4 and row 5 to row 3, the 6×6 secular equation in Eq. (2) can be reduced to a 4×4 one:

$$
\begin{vmatrix}
x_1 - \gamma & x_1 & x_1 & x_1 \\
x_2 & x_2 y_1 - \gamma & x_2 y_{12} & x_2 y_1 y_{12} \\
x_3 + x_5 & (x_3 + x_5) y_{21} & (x_3 + x_5) y_2 - \gamma^3 & (x_3 + x_5) y_2 y_{21} \\
x_4 + x_6 & (x_4 + x_6) y_{21} & (x_4 + x_6) y_2 & (x_4 + x_6) y_2 y_{21} - \gamma^3
\end{vmatrix} = 0
\tag{3}
$$

This is the general secular equation for this mosaic multiple-binding model with all four cooperativity parameters included. Further simplification can be made if some of the cooperativity parameters have a value of one (no cooperativity).

Step 5: Binding Isotherms

The fractions of biopolymer subunits bound with ligand I and ligand II, θ_I and θ_{II}, as a function of the concentrations of the two ligands in solution can be evaluated using the relations:

$$\theta_I = \frac{\partial \ln \gamma_1}{\partial \ln X_I}, \qquad \theta_{II} = \frac{\partial \ln \gamma_1}{\partial \ln X_{II}} \tag{4}$$

where γ_1 is the largest root of the secular equation in Eq. (3). To calculate Eq. (4), the weights $x_1, x_2, \ldots$ in Eq. (3) must be expressed in terms of X_I and X_{II} by using the relations in Fig. 2. On the other hand, one can also calculate the probabilities of unit 1, unit 2, and so on, using the equations

$$\theta_i = \frac{\partial \ln \gamma_1}{\partial \ln x_i}, \qquad i = 1, 2, \ldots, 6 \tag{5}$$

Then, θ_I and θ_{II} can be evaluated using the relations

$$\begin{aligned} \theta_I &= \theta_2 + \theta_4 + \theta_5 + 2\theta_6 \\ \theta_{II} &= \theta_3 + \theta_4 + \theta_5 + \theta_6 \end{aligned} \tag{6}$$

Discussion

The present formalism has two limitations. First, the formalism is derived on the basis of the assumption that the biopolymer in the binding system is infinitely long. Therefore, the formalism is not applicable if the biopolymer is too short. Second, the formalism is applicable only when the number of the elementary units of the system is finite. Thus, it may not work for some multiple-binding systems (see Chen[13]). On the other hand, the formalism always works for single-binding systems because the number of ligand species in the system is finite. That is, complex ligand-binding systems, such as those involving ligand polymerizations or the piggy-back binding mechanism,[9,16] can all be analyzed by the present formalism, as long as each subunit of the biopolymer contains only one binding site.

As can be seen from steps 3–5, the binding of two ligands in this multiple-binding model is equivalent to the binding of six ligand species (defined by the six elementary units) to a biopolymer with single-binding sites. That is, when the elementary units and their statistical weights of an arbitrary binding system (single binding and multiple binding alike) are

[16] S. Campbell and A. Maxwell, *J. Mol. Biol.* **320,** 171 (2002).

identified, the evaluation of the secular equation and the binding isotherms can just follow the usual "matrix" method used in linear Ising problems, if adjustments due to ligand–ligand interactions in evaluating the statistical weight of some units and the n-mer effect (caused by ligands that cover more than one subunit) in obtaining the secular equation are properly made (see above).

[9] Probing Site-Specific Energetics in Proteins and Nucleic Acids by Hydrogen Exchange and Nuclear Magnetic Resonance Spectroscopy

By IRINA M. RUSSU

Introduction

The macromolecular structure of a protein or a nucleic acid results from an array of interactions between its building blocks and its structural elements. These interactions act in concert to ensure the thermodynamic and conformational accuracy of the final molecular architecture. An important question in understanding these structures is how, and how much, each structural element contributes to the stability of the final product. In addition to its fundamental relevance, the question is of practical interest for predicting effects of site-directed mutagenesis on the molecule of interest. The question of site-specific energetics is also essential for understanding the mechanisms by which the molecule performs its function. All processes involved in biological function, such as formation of macromolecular assemblies, binding of ligands, and enzymatic catalysis, are directed by pathways of energetic couplings between individual sites. Understanding how a molecule works necessitates mapping these pathways and quantifying their energetic contents. The present chapter discusses one methodology currently used to address the question of site-specific energetics in biological molecules. The methodology combines hydrogen exchange and nuclear magnetic resonance (NMR) spectroscopy.

Hydrogen exchange has been extensively used to elucidate structural dynamics of proteins and nucleic acids. Excellent reviews of this work are available, including those in this and previous volumes of this series.[1–4]

[1] S. W. Englander and J. J. Englander, *Methods Enzymol.* **232**, 26 (1994).
[2] M. Gueron and J. L. Leroy, *Methods Enzymol.* **261**, 383 (1995).

In proteins, investigations have focused on the exchange of peptide NH hydrogens located in the protein backbone. Similar studies of hydrogens from amino acid side chains have been less extensive.[5-7] This was due, in part, to technical difficulties in distinguishing these hydrogens, and their exchange, from peptide hydrogens. Advances in spectroscopic techniques, such as Raman and NMR, now allow direct observation of side-chain hydrogens and measurements of their exchange with solvent hydrogens under wide ranges of experimental conditions. The present chapter illustrates hydrogen exchange studies of proteins by specific examples using side-chain hydrogens. In addition, the chapter also includes examples of hydrogen exchange studies of nucleic acids by NMR. This inclusion is timely because of a number of emerging studies of protein–nucleic acid complexes by hydrogen exchange.[8,9] It is also hoped that the discussion will highlight parallels and differences in hydrogen exchange processes between proteins and nucleic acids which may be useful for future investigations of these complexes.

Mechanism of Hydrogen Exchange

The current model for exchange of hydrogens in biological molecules postulates a conformational transition between two states, commonly labeled "closed" and "open"[10]:

$$(H)_{closed} \underset{k_{cl}}{\overset{k_{op}}{\rightleftharpoons}} (H)_{open} \overset{k_{ex,\,open}}{\longrightarrow} (H)_{exchanged}$$

In the closed state, the hydrogen cannot exchange because of its participation in hydrogen bonding and lack of accessibility to water and exchange catalysts. The conformational transition results in an open state in which barriers to exchange are removed and the hydrogen becomes exchange competent. The nature of the structural transition(s) between closed and open states is not yet fully understood. In proteins, under native conditions, exchange-competent states result from changes in structure ranging from local fluctuations to large conformational transitions, including cooperative

---

[3] Y. Bai, J. J. Englander, L. Mayne, J. S. Milne, and S. W. Englander, *Methods Enzymol.* **259,** 344 (1995).

[4] C. B. Arrington and A. D. Robertson, *Methods Enzymol.* **323,** 104 (2000).

[5] E. Tuchsen and C. Woodward, *Biochemistry* **26,** 8073 (1987).

[6] P. Rajagopal, B. E. Jones, and R. E. Klevit, *J. Biomol. NMR* **11,** 205 (1998).

[7] G. P. Connelly and L. P. McIntosh, *Biochemistry* **37,** 1810 (1998).

[8] G. M. Dhavan, J. Lapham, S. Yang, and D. M. Crothers, *J. Mol. Biol.* **288,** 659 (1999).

[9] C. G. Kalodimos, R. Boelens, and R. Kaptein, *Nat. Struct. Biol.* **9,** 193 (2002).

[10] A. Hvidt and S. O. Nielsen, *Adv. Protein Chem.* **21,** 287 (1966).

global unfolding.[3,11-13] In nucleic acids, the transition yielding exchange-competent states for imino hydrogens is generally equated to the opening of the base pair.[2,13]

The kinetics of the exchange process observed experimentally depends on the rate constants of the opening and closing reactions (k_{op} and k_{cl}, respectively) and on the rate of exchange from the open state ($k_{ex, open}$). For native structures, thermodynamic stability dictates that $k_{cl} \gg k_{op}$. In this case, the observable exchange rate is[10,13]

$$k_{ex} = \frac{k_{op} \cdot k_{ex, open}}{k_{op} + k_{cl} + k_{ex, open}} \tag{1}$$

In practice, further simplifications of Eq. (1) are commonly used. Two limiting cases are distinguished depending on how the rate of exchange from the open state compares with the rate of closing.

EX2 (Bimolecular Exchange) Regime

This regime occurs when the exchange from the open state is slow (i.e., $k_{ex, open} \ll k_{cl}$) and rate limiting. Then

$$k_{ex} = K_{op} \cdot k_{ex, open} \tag{2}$$

where $K_{op} = k_{op}/k_{cl}$ is the equilibrium constant of the opening reaction.

EX1 (Unimolecular Exchange) Regime

In the EX1 regime, the transfer of the hydrogen from the open state is fast (i.e., $k_{ex, open} \gg k_{cl}$), and exchange occurs at every opening event. Then

$$k_{ex} = k_{op} \tag{3}$$

The exchange of the hydrogen from the open state is generally independent of the structural opening reaction. The exchange is assumed to occur in a manner similar to the exchange of a chemically identical hydrogen in short peptides or in mononucleotides. The exchange is catalyzed by H^+, OH^-, water, and hydrogen acceptors present in the solvent. Accordingly, the rate of exchange from the open state is

$$k_{ex, open} = k_w + k_H \cdot [H^+] + k_{OH} \cdot [OH^-] + \sum_i k_B^{(i)} \cdot [B]_i \tag{4}$$

[11] G. Wagner and K. Wüthrich, *J. Mol. Biol.* **134,** 75 (1979).
[12] C. Woodward, I. Simon, and E. Tuchsen, *Mol. Cell. Biochem.* **48,** 135 (1982).
[13] S. W. Englander and N. R. Kallenbach, *Q. Rev. Biophys.* **16,** 521 (1984).

where k_H and k_{OH} are the rate constants for H^+ and OH^- catalysis, respectively, $k_B^{(i)}$ is the rate constant for catalysis by catalyst B_i, k_w is the rate of water-catalyzed exchange, and the sum includes all catalysts B_i present in solution (other than H^+ and OH^-). The dependence of the exchange rate on the concentration of exchange catalysts provides a direct way to distinguish between the EX1 and the EX2 regime. In the EX2 regime, an increase in catalyst concentration results in an enhancement of the exchange rate [Eqs. (2) and (4)]. In contrast, in the EX1 regime, the exchange is determined solely by the structural opening step [Eq. (3)] and is independent of catalyst concentration. A lack of dependence on catalyst concentration may also be observed when the exchange is efficiently catalyzed by internal catalysis, namely, by transfer of the hydrogen to groups within the same molecule. This kind of catalysis has been observed in nucleic acids, where imino hydrogens can be transferred to nitrogen atoms of the bases lacking imino groups, that is, adenine and cytosines. Experimental evidence available thus far suggests that the structural opening reaction involved in this internal catalysis is the same as that operating in catalysis by external hydrogen acceptors.[14]

The use of exchange data for short peptides or mononucleotides to interpret hydrogen exchange measurements can be limited by effects of macromolecular structure on the rate of hydrogen transfer from the open state. For example, the geometry of the open state may lower the transfer rate by partially blocking access of the catalyst to the exchangeable hydrogen. More complex effects originate from the electrostatic potential created by the macromolecular structure at the site of the exchanging hydrogen. These effects produce alterations in the rate constants for charged catalysts, for example, k_H and k_{OH} in Eq. (4), and the alterations are pH dependent. As a result, the exchange reactions deviate from the first-order catalysis observed in model compounds.[15]

NMR Methods for Measuring Hydrogen Exchange Rates

Hydrogen exchange rates in proteins and nucleic acids encompass more than 10 orders of magnitude, namely, from less than 10^{-8} s^{-1} (exchange times longer then 1 year) to $\sim 10^2$ s^{-1} or faster (exchange times of milliseconds or less). Among these, NMR methods can access only two ranges. On the one hand, hydrogen–deuterium (H–D) exchange measurements permit observation of rates slower than 10^{-2}–10^{-3} s^{-1} (depending on the

[14] M. Gueron, M. Kochoyan, and J. L. Leroy, *Nature* **328,** 89 (1987).
[15] M. Delepierre, C. M. Dobson, M. Karplus, F. M. Poulsen, D. J. States, and R. E. Wedin, *J. Mol. Biol.* **197,** 111 (1987).

experimental setup as described below). On the other hand, experiments of transfer of magnetization from water can measure rates of exchange faster than ~ 0.1–0.5 s^{-1}, up to ~ 100 s^{-1}. This range can be extended to ~ 1000 s^{-1} by using indirect NMR methods such as linewidth measurements. When the exchange is too fast to be detectable by H–D exchange and too slow for transfer of magnetization experiments, one can attempt to modify experimental conditions (such as temperature and pH) to speed up, or to slow down, exchange such that it becomes observable by NMR. This approach is often limited by the range of experimental conditions in which the integrity of the molecule is preserved.

The use of NMR spectroscopy to monitor hydrogen exchange requires prior assignments of resonances to individual protons in the molecule. The methods for obtaining these assignments are well established for proteins and nucleic acids, and have been described extensively elsewhere.[16,17]

Hydrogen–Deuterium Exchange NMR Measurements

The protocol in these experiments is similar to that used in hydrogen–tritium exchange. At the beginning of the measurement a protonated sample is rapidly switched into a deuterated solvent. This can be accomplished by diluting a concentrated solution or a lyophilized sample into the buffer in D_2O. Spectra are recorded at regular intervals after initiation of exchange. Depending on the resolution of the exchangeable proton resonances of interest in the spectrum one uses for observation one-dimensional (1D) ^1H NMR pulse sequences, such as Jump-and-Return[18] and WATERGATE,[19] or two-dimensional (2D) pulse sequences, such as ^1H–^1H correlation spectroscopy (COSY) and total correlation spectroscopy (TOCSY), and ^1H–^{15}N heteronuclear single quantum coherence (HSQC).[17,20,21] The use of heteronuclear ^1H–^{15}N 2D pulse sequences requires samples labeled with ^{15}N, uniformly or at specific amino acid residues. The intensity of each resonance or 2D cross-peak is measured for each spectrum, using integration software packages, generally provided by NMR manufacturers. The final data are fitted as a function of the exchange delay τ to the equation

[16] K. Wüthrich, "NMR of Proteins and Nucleic Acids." John Wiley & Sons, New York, 1986.
[17] J. Cavanagh, W. J. Fairbrother, A. G. I. Palmer, and N. J. Skelton, "Protein NMR: Principles and Practice." Academic Press, San Diego, CA, 1996.
[18] P. Plateau and M. Gueron, *J. Am. Chem. Soc.* **104**, 7310 (1982).
[19] M. Piotto, V. Saudek, and V. Sklenar, *J. Biomol. NMR* **2**, 661 (1992).
[20] S. Grzesiek and A. Bax, *J. Am. Chem. Soc.* **115**, 12593 (1993).
[21] S. Mori, C. Abeygunawardana, M. O. Johnson, and P. C. M. van Zijl, *J. Magn. Reson. B* **108**, 94 (1995).

$$I(\tau) = [I(0) - I(\infty)] \cdot e^{-k_{ex} \cdot \tau} + I(\infty) \qquad (5)$$

where $I(\tau)$ is the intensity of the resonance (cross-peak) at exchange time τ, $I(0)$ is the intensity at $\tau = 0$, $I(\infty)$ is the intensity in the fully exchanged sample, and k_{ex} is the exchange rate. The intensity $I(\infty)$ differs from zero when residual H_2O is present in the sample. The fastest exchange rate that can be measured in these experiments is determined by the time elapsed between the initiation of exchange and the first NMR observation. This "dead" time is required for temperature equilibration and for setting up the NMR spectrometer at correct acquisition parameters, and is expected to vary from user to user. To shorten this time it is recommended that the homogeneity of the magnet be adjusted on a sample of the same volume containing only buffer. Moreover, the deuterated buffer used to initiate the exchange can be separately equilibrated at the desired temperature. Several experimental setups have been designed to allow initiation of the exchange inside the NMR probe and, thus, to shorten the time before the NMR experiment to less than 1 min (e.g., Refs. 2 and 22). The time resolution of each experiment depends on the time required for the acquisition of each spectrum. This time, in turn, depends on the method used (e.g., 1D or 2D pulse sequence) and on the concentration of the sample. For a typical NMR sample ($\sim$1–2 mM), a spectrum of satisfactory signal-to-noise ratio can be acquired in a time as short as a few minutes.

Transfer of Magnetization from Water Experiments

The sample is prepared in aqueous solvents and the exchange is initiated by perturbing the magnetization of water protons. During the exchange delay τ that follows, this perturbation is transmitted to all exchangeable hydrogens in the molecule via the exchange process. In the simplest version of the experiment, the initial perturbation may consist of inversion of the water proton resonance, using a selective pulse. After the exchange delay, the spectrum is acquired using 1D ^1H or 2D ^1H–^{15}N HSQC pulse sequences, either as a regular spectrum or as a difference spectrum, with and without water inversion.[23,24]

In a 1D ^1H NMR spectrum, the dependence of the intensity of an exchangeable proton resonance on the exchange delay τ is given by

$$I(\tau) = I^0 + [I(0) - I^0 - A] \cdot e^{-(R_1 + k_{ex}) \cdot \tau} + A \cdot e^{-R_{1w} \cdot \tau} \qquad (6)$$

[22] A. K. Bhuyan and J. B. Udgaonkar, *Proteins Struct. Funct. Genet.* **30**, 295 (1998).
[23] S. Grzesiek and A. Bax, *J. Biomol. NMR* **3**, 627 (1993).
[24] S. Mori, C. Abeygunawardana, P. C. M. van Zijl, and J. M. Berg, *J. Magn. Reson. B* **110**, 96 (1996).

where I^0 is the intensity at equilibrium, $I(0)$ is the intensity immediately after the selective pulse on water, k_{ex} is the exchange rate, R_1 is the longitudinal relaxation rate of the observed proton, and R_{1w} is the longitudinal relaxation rate of water protons. The factor A is defined as

$$A = \left[\frac{I_w(0)}{I_w^0} - 1 \right] \cdot \frac{k_{ex}}{R_1 + k_{ex} - R_{1w}} \cdot I^0 \tag{7}$$

in which $I_w(0)$ and I_w^0 are the intensities of the water proton resonance after the inversion pulse and at equilibrium, respectively. The term $I(0)$ in Eq. (6) accounts for the effects of exchange during the selective pulse used to invert the water proton magnetization. In general, $I(0) < I^0$. Because the length of the selective pulse is in the range of milliseconds, these effects are significant only when the exchange is fast (e.g., when k_{ex} is $\sim 50 \text{ s}^{-1}$ for a 6-ms pulse). When the exchange is slower, $I(0) \approx I^0$ and Eq. (6) can be simplified as

$$I(\tau) = I^0 - A \cdot [e^{-(R_1 + k_{ex})\tau} - e^{-R_{1w} \cdot \tau}] \tag{8}$$

Determination of the exchange rate k_{ex} from the dependence of $I(\tau)$ on τ [Eq. (6) or (8)] requires independent knowledge of $I_w(0)/I_w^0$, R_{1w}, and R_1. $I_w(0)/I_w^0$ expresses the magnitude of the perturbation (e.g., inversion) of the water proton resonance by the selective pulse. This ratio is determined in a separate experiment with the same acquisition parameters as those used in magnetization transfer experiments except that a single low-power pulse is used for observation of water protons. The longitudinal relaxation rate of water protons (R_{1w}) is similarly measured by an inversion–recovery pulse sequence.[25] The longitudinal relaxation rate of the proton of interest (R_1) cannot be readily measured. Instead, the sum $R_1 + k_{ex}$ is obtained by fitting the intensity $I(\tau)$ as a function of the exchange delay τ to Eq. (6) or (8), or is measured using, for example, a selective saturation–recovery pulse sequence.[25] The exchange rate is then calculated from the factor A, which is also obtained from the fit.

Equations (6) and (8) represent first approximations of the time dependence of the resonance intensity in magnetization transfer experiments. They are valid for a single proton in exchange with water. Biological molecules are in fact multi spin proton systems. Thus, the time dependence of the magnetization can be greatly affected by transfers of magnetization to neighboring protons [i.e., cross-relaxation or nuclear Overhauser effect (NOE)]. Two cross-relaxation pathways have been shown to be important

[25] R. Freeman, "A Handbook of Nuclear Magnetic Resonance Spectroscopy." Longman Scientifical & Technical, Essex, UK, 1988.

in transfer of magnetization experiments. The first consists of cross-relaxation between the observed proton and neighboring protons, which are exchanging fast (e.g., hydroxyl or amine groups). The second involves cross-relaxation with non exchangeable protons, and are especially important in measurements of exchange of backbone amide hydrogens in proteins. This is because, in proteins, $C_\alpha H$ resonances occur close to, or overlap with, the water proton resonance. As a result, inversion of the water resonance also perturbs $C_\alpha H$ resonances. For large molecules in the slow-motion limit, effects of cross-relaxation can be removed by using mixing schemes that combine nuclear Overhauser effect spectroscopy (NOESY) and rotating frame Overhauser effect spectroscopy (ROESY) mixing schemes such as CLEANEX-PM spin-locking pulse sequence.[26] Cross-relaxation with $C_\alpha H$ protons can be effectively suppressed by purging $C_\alpha H$ resonances using ^{13}C pulses.[23] The experiments require samples uniformly labeled with [13]C. Further reduction of the effects of cross-relaxation can be obtained by carrying out the transfer of magnetization experiments for short exchange times.[24] In this initial rate approximation, the dependence of the resonance intensity on the exchange delay is linear:

$$I(\tau) = I(0) - \left\{ [I(0) - I^0] \cdot (R_1 + k_{ex}) - \left[\frac{I_w(0)}{I_w^0} - 1 \right] \cdot k_{ex} \cdot I^0 \right\} \cdot \tau \qquad (9)$$

When exchange is not too fast, such that $I(0) \approx I^0$, the exchange rate can be readily calculated from the slope of this linear dependence.

One special precaution in experiments of transfer of magnetization from water concerns radiation damping. Radiation damping affects the time evolution of the water proton magnetization, especially for the state-of-the-art NMR spectrometers currently in use. In the presence of a residual transverse magnetization, the water magnetization returns to equilibrium at a rate that can be much faster than its intrinsic relaxation rate (R_{1w}). As shown in Eqs. (6) and (8), this faster recovery would affect the exchange rate measurement. The effects of radiation damping can be greatly reduced by applying a weak gradient (e.g., <1 G/cm) during the exchange delay.

Resonance Linewidth

The linewidth of an exchangeable proton resonance at half-height ($\Delta\nu_{1/2}$) is related to the exchange rate k_{ex} by[27]

$$\pi \cdot \Delta\nu_{1/2} = R_2 + k_{ex} \qquad (10)$$

[26] T.-L. Hwang, P. C. M. van Zijl, and S. Mori, *J. Biomol. NMR* **11**, 221 (1998).
[27] J. Sandstrom, "Dynamic NMR Spectroscopy." Academic Press, London, 1982.

where R_2 is the transverse relaxation rate of the proton. This relationship has been extensively used for measurements of exchange rates. One limitation of this approach is that determination of the exchange rate requires independent knowledge of the R_2 rate. For this reason, linewidth measurements have been commonly used to look for changes in exchange rates. Increases in linewidth between 10 and 300 Hz, which can be readily detected experimentally, correspond to changes in exchange rates of $\sim$30 to $\sim$1000 s^{-1}. Reliable measurements of the exchange rates also require verification that the experimental variable that induces these changes does not affect also the relaxation rate R_2. For example, as described in the previous section, it is often of interest to measure hydrogen exchange rates as a function of the concentration of catalyst in order to approach the EX1 regime [Eqs. (1)–(4)]. Increasing concentrations of catalyst enhance exchange rates but may also affect proton relaxation. Our laboratory has previously shown that this is the case in DNA double helices, where increasing concentrations of exchange catalyst (ammonia) affect the transverse relaxation rates of both exchangeable and nonexchangeable protons, most likely by aggregation of the DNA.[28] In situations such as this, linewidth measurements cannot be used reliably to determine exchange rates.

Probing Site-Specific Energetics by Hydrogen Exchange

Evaluation of site-specific energetics by hydrogen exchange follows directly from the structural opening model described in a previous section. According to this model, the exchange rate of a hydrogen is directly related to the energetic cost of the structural opening reaction. The free energy change in the opening reaction is

$$\Delta G_{op} = -RT \ln K_{op} \qquad (11)$$

where R is the gas constant and T is the absolute temperature. This free energy change represents a measure of the stability of molecular structure at the site of interest: the larger the free energy for opening ΔG_{op}, the higher the structural stability at the corresponding site. Definition of this structural stability in molecular terms is possible only to the extent that the conformational transitions constituting the opening reaction are known. The opening reaction is defined operationally as the reaction leading to an exchange-competent state. For exchange of imino hydrogens in nucleic acids, the reaction is thought to consist of the opening of the base

---

[28] E. Folta-Stogniew and I. M. Russu, *Biochemistry* **35**, 8439 (1996).

pair. Strictly speaking, the exchange requires only the opening of the base that contains the imino proton (i.e., guanine, thymine, or uracil). The process is pictured as swinging this base out of the structure, into a solvent-accessible state (e.g., in double helices, a rotation of the base into the major or the minor groove by 50° or more[29]). Molecular dynamics simulations show, however, that this opening reaction is accompanied by significant perturbations of the partner base.[29] In proteins, the conformational transitions yielding exchange vary depending on the observed hydrogen and on experimental conditions; for example, exchange may occur by global unfolding, by limited unfolding of units of structure, or by small local fluctuations. Under native conditions, many hydrogens in stable proteins exchange via local fluctuations.[30,31] For a local fluctuation to be effective in exchange, any hydrogen bond to the exchanging hydrogen must break such that a new, transient hydrogen bond to the exchange catalyst can form. Breaking of the hydrogen bond is necessary but not sufficient. The local fluctuation should also bring the exchanging hydrogen into such a state that the hydrogen is exposed to the catalyst, or the catalyst itself can approach it through internal diffusion. It is therefore expected that formation of this "open" state should require some distortion of the structure around the exchanging hydrogen. The energetic cost of this transient structural perturbation adds to the cost of breaking the hydrogen bond to give the overall ΔG_{op} value obtained from the experiment.

The applications of hydrogen exchange techniques covered in this chapter are aimed at assessing how changes in structure affect structural energetics at individual sites. Consider a change in structure resulting from a functional transition, binding of a ligand, or a single modification such as substitution of an amino acid in a protein, or of a nucleotide in a nucleic acid. The effect of this change on the stabilization free energy at any site in the macromolecule can be expressed as[13]

$$\delta\Delta G_{op} = \Delta G_{op} - \Delta G'_{op} = -RT \ln \frac{K_{op}}{K'_{op}} \tag{12}$$

where the prime symbols refer to the equilibrium constant and the free energy change in the opening reaction at the observed site in the presence of the structural change. The strategies used to obtain the site-specific energetic cost of the structural change, $\delta\Delta G_{op}$, depend on the regime of exchange for the observed hydrogen.

[29] P. Varnai and R. Lavery, *J. Am. Chem. Soc.* **124,** 7272 (2002).

[30] J. S. Milne, L. Mayne, H. Roder, A. J. Wand, and S. W. Englander, *Protein Sci.* **7,** 739 (1998).

[31] H. Maity, W. K. Lim, J. N. Rumbley, and S. W. Englander, *Protein Sci.* **12,** 153 (2003).

If the exchange is in the EX2 regime, the variation in exchange rate can be directly related to the change in the equilibrium constant K_{op} [Eq. (2)] and to the corresponding $\delta\Delta G_{op}$:

$$\frac{k_{ex}}{k'_{ex}} = \frac{K_{op}}{K'_{op}} \quad \text{and} \quad \delta\Delta G_{op} = -RT \cdot \ln\frac{k_{ex}}{k'_{ex}} \tag{13}$$

It is important to note that the underlying assumption in Eq. (13) is that the change in structure does not affect the rate of exchange from the open state ($k_{ex, \, open}$) at the observed site [Eq. (2)]. Because the open state is not directly accessible experimentally, establishing the validity of this assumption is often difficult.

If exchange data are available for the EX1 regime, one can estimate the change in the equilibrium constant K_{op} from the change in the rate of the opening reaction (k_{op}). This evaluation is possible by making the assumption that the change in structure does not affect the rate of closing (k_{cl}). In this case,

$$\frac{K_{op}}{K'_{op}} = \frac{k_{op}}{k'_{op}} \quad \text{and} \quad \delta\Delta G_{op} = -RT \cdot \ln\frac{k_{op}}{k'_{op}} \tag{14}$$

The following sections illustrate how these approaches have been used to (1) map the energetic effects of single-site substitutions in a DNA structure, and (2) trace the energetic pathways of the structural transition that occurs in human hemoglobin on oxygen binding.

Mapping Energetic Effects of Single-Site Substitutions

The molecules discussed in this illustration are the intramolecular DNA triple helices (triplexes) shown in Fig. 1. The triplex labeled C⁺·GC contains only canonical triads, namely, four C⁺·GC triads and three T·AT triads. In these triads, the base in the third strand (C⁺ and T, respectively) binds to the purine in the corresponding Watson–Crick base pair (G in GC and A in AT) by two Hoogsteen hydrogen bonds. In the other two triplexes, the canonical C⁺18·G11C4 triad is replaced, respectively, by T·CG and G·TA triads. These two substituted triplexes are of interest for two reasons. First, one notes that, in T·CG and G·TA triads, the Hoogsteen hydrogen bonding from the base in the third strand (T and G, respectively) occurs at the pyrimidines in the double-helical part of the structure (C and T, respectively). Therefore, whereas canonical C⁺·GC and T·AT triads recognize only purines in double-helical DNA, the noncanonical G·TA and T·CG triads extend this recognition code to pyrimidines. This extension of the recognition code is of great interest in antigene strategies to control

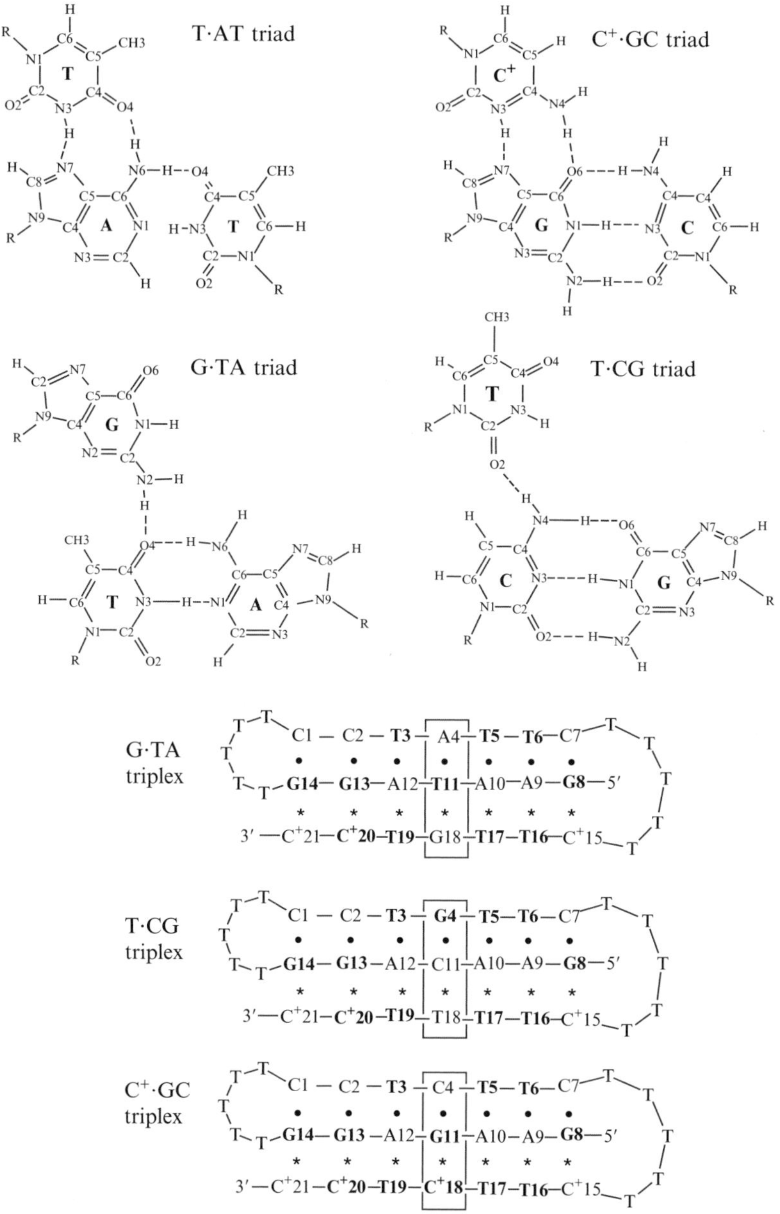

FIG. 1. Base triads and intramolecular folding of the three DNA triple helices. In each triple helix, dots indicate Watson–Crick hydrogen bonding and asterisks indicate Hoogsteen hydrogen bonding. The resonances of imino protons from the bases shown in boldface in each triple helix are observable in NMR spectra.

gene expression. Second, from a structural point of view, the G·TA and T·CG triplexes offer a test case to assess the relationship between structure and structural energetics in nucleic acids. The structures of the two triplexes, solved using NMR methods,[32] differ largely from that of the canonical C⁺·GC triplex. However, the two substituted triplexes are similar to each other, indicating that the magnitude and pattern of structural perturbations induced by the T·CG triad are the same as those induced by the G·TA triad. In spite of these structural similarities, the global thermodynamic stability of the T·CG triplex (measured, e.g., by optical melting) is much lower than that of the G·TA triplex. To elucidate the molecular origin of this difference in global stability we have probed the energetics of individual triads in these triplexes, using the exchange of imino hydrogens in each Watson–Crick and Hoogsteen base pair in the structures.[33] The resonances of imino protons in C⁺·GC triplex (i.e., G-N1H, T-N3H, and C⁺-N3H) and their assignments[33] are shown in Fig. 2A. Figure 2 also illustrates the time dependence of these resonances in the transfer of magnetization from water (Fig. 2B) and in H–D exchange (Fig. 2C) experiments. Examples of the exchange curves obtained in these experiments are shown in Fig. 3. The exchange rates for all imino hydrogens in C⁺·GC triplex are summarized in Table I. They range from 3×10^{-4} s^{-1} for T17 to 45 s^{-1} for C⁺20. One notes that for two hydrogens (i.e., T6 and G11) the exchange is too fast to be measured in H–D exchange experiments (i.e., $k_{ex} > 4 \times 10^{-3}$ s^{-1} in our experimental setup), and too slow to be measured in transfer of magnetization experiments (i.e., $k_{ex} < 0.3$ s^{-1}). In general, one would attempt to modify experimental conditions such that the exchange rates of these hydrogens become measurable by NMR. For the C⁺·GC triplex this approach has not been successful because of the narrow ranges of temperature and pH in which the triple-helical structure is stable (i.e., temperature below 15° and pH < 6.0).[33]

The effects of C⁺·GC → G·TA and C⁺·GC → T·CG triad substitutions on the energetics of the structure depend on the nature of the substitution and on the location of the observed site. For example, as shown in Fig. 3A for the first triad, C⁺15·G8C7, the exchange of the imino hydrogen in the Watson–Crick base pair is indistinguishable between the three triplexes. In contrast, for the T19·A12T3 triad located next to the substitution site, the exchange rate of T19 imino hydrogen increases 2.5-fold in the G·TA triplex and 350-fold in the T·CG triplex (relative to C⁺·GC

[32] I. Radhakrishnan and D. J. Patel, *Biochemistry* **33**, 11405 (1994).
[33] D. Coman and I. M. Russu, *Biochemistry* **41**, 4407 (2002).

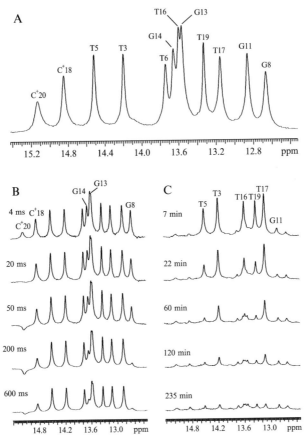

Fig. 2. (A) Imino proton resonances in the control NMR spectrum of $C^+ \cdot GC$ triplex in 90% H_2O–10% D_2O at pH 4.6 and 5°. (B) Selected spectra from experiments of transfer of magnetization from water. (C) Selected spectra from H–D exchange experiments. In (B) and (C), the exchange delay is given for each spectrum, and the imino proton resonances that show changes in intensity in each experiment are indicated.

triplex). These changes in exchange rate can be expressed as variations in free energy of structural stabilization based on Eq. (13). The use of this equation is justified by the fact that, under the experimental conditions used, the exchange of imino hydrogens (except those in C^+) is in the EX2 regime.[34] The changes in structural stabilization free energy in each

[34] S. W. Powell, L. Jiang, and I. M. Russu, *Biochemistry* **40**, 11065 (2001).

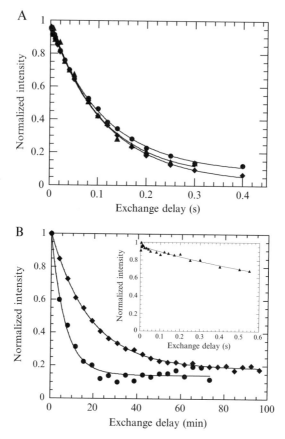

FIG. 3. Exchange curves for imino hydrogens of G8 (A) and of T19 (B) in $C^+ \cdot GC$ triplex (♦), G·TA triplex (●), and T·GA triplex (▲). The exchange of G8 imino hydrogen was measured in transfer of magnetization experiments and the exchange rates are as follows: 4.2 ± 0.1 s^{-1} in $C^+ \cdot GC$ triplex, 4.1 ± 0.1 s^{-1} in G·TA triplex, and 4.6 ± 0.3 s^{-1} in T·CG triplex. The exchange of T19 imino hydrogen in $C^+ \cdot GC$ and G·TA triplexes was measured in H–D exchange experiments and the exchange rates are $(9.9 \pm 0.9) \times 10^{-4} s^{-1}$ and $(2.5 \pm 0.3) \times 10^{-3} s^{-1}$, respectively. The exchange of T19 imino hydrogen in T·CG triplex [*inset* in (B)] was measured in transfer of magnetization experiments and the exchange rate is 0.35 ± 0.01 s^{-1}.

substituted triplex, relative to $C^+ \cdot GC$ triplex, at 5° are summarized in Fig. 4. These energetics maps identify the base pairs whose stabilities are affected by the single triad substitutions. They also provide an excellent illustration of the complementarities between structure and single-site energetics. Although the two triplexes are structurally isomorphous, the

TABLE I

EXCHANGE RATES[a] OF IMINO HYDROGENS IN C$^+$·GC TRIPLEX IN 100 mM NaCl AND 5 mM MgCl$_2$ AT pH 4.6 AND 5°

C1	C2	T3	C4	T5	T6	C7
		$(4.1 \pm 0.3) \times 10^{-4}$		$(1.2 \pm 0.1) \times 10^{-3}$	$4 \times 10^{-3} < k_{ex} < 0.3$	
G14	G13	A12	G11	A10	A9	G8-5'
1.69 ± 0.06	0.43 ± 0.01		$4 \times 10^{-3} < k_{ex} < 0.3$			4.2 ± 0.1
3'-C$^+$21	C$^+$20	T19	C$^+$18	T17	T16	C$^+$15
b	45 ± 2	$(9.9 \pm 0.9) \times 10^{-4}$	2.4 ± 0.1	$(3.0 \pm 0.3) \times 10^{-4}$	$(0.9 \pm 0.1) \times 10^{-3}$	b

[a] In s^{-1}.
[b] Resonance is broadened beyond detection because of fast exchange of the imino hydrogen [Eq. (10)].

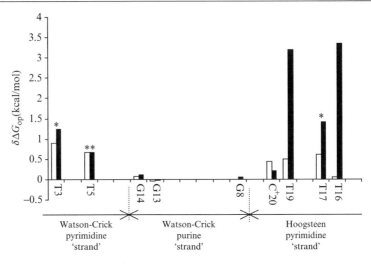

FIG. 4. Site-specific energetic effects of the triad substitutions in G·TA triplex (open columns) and in T·CG triplex (solid columns). $\delta\Delta G_{op}$ (at $5°$) is calculated on the basis of Eq. (13). For the bases labeled with an asterisk (*), the triad substitution increases the exchange rate of the imino hydrogen into the range 4×10^{-3}–0.3 s^{-1}. The bars shown are calculated assuming the smallest change in exchange rate (i.e., $k_{ex} = 4 \times 10^{-3}$ s^{-1}).

perturbations in energetics at some sites in the T·CG triplex are higher than those in the G·TA triplex.

Structural Energetics of Functional Transition in Human Hemoglobin

Human hemoglobin (Hb) undergoes a transition on binding of ligand, from a low-affinity deoxy-T to a high-affinity ligated-R structure. The transition affects the tertiary structures of each α and β subunit as well as the interfaces between subunits. The changes in quaternary structure are localized mostly at the interfaces between the $\alpha\beta$ dimers (i.e., $\alpha_1\beta_2$ interface and the symmetry-related $\alpha_2\beta_1$ interface), and at the contacts between α subunits (i.e., $\alpha_1 \alpha_2$). In contrast, the interfaces between subunits within each $\alpha\beta$ dimer (i.e., $\alpha_1\beta_1$ and $\alpha_2\beta_2$ interfaces) do not undergo significant changes.[35]

[35] M. F. Perutz, *Nature* **228,** 726 (1970).

The Hb tetramer binds the first ligand with low binding energy. The binding changes the structural free energy within the molecule such that the energy of binding ligand at the other heme sites is enhanced. The pathways for this transmission of energy between hemes have been analyzed structurally by X-ray crystallography[35,36] and by NMR,[37] and energetically by the thermodynamics of oxygenation and dimer–tetramer equilibria[38] and by hydrogen exchange. The use of hydrogen exchange to probe the allosteric transition in Hb has been initiated and developed by Englander and co-workers. Their studies have concentrated on the exchange of peptide NH hydrogens. Using hydrogen–tritium exchange, these authors have demonstrated that $\sim$25% of the peptide NH hydrogens in Hb change their exchange rates on ligand binding. The exchange rates of these hydrogens are faster in the ligated-R than in the deoxy-T structure by factors ranging from 15 to 10,000.[1] Several of these allosterically sensitive hydrogens have been localized in the Hb structure near the amino termini of α subunits at the $\alpha_1\alpha_2$ contact, and at the $\alpha_1\beta_2/\alpha_2\beta_1$ interfaces (in the F helix and the FG segment, and at the carboxyl termini of the β subunits).[1] Our laboratory has extended these investigations and has identified several new allosterically sensitive sites, using NMR spectroscopy. The following illustrates the methodology, using results for two sites: Trpβ37(C3) at the interdimeric $\alpha_1\beta_2/\alpha_2\beta_1$ interfaces, and Hisα122(H5) at the intradimeric $\alpha_1\beta_1/\alpha_2\beta_2$ interfaces.[39,40]

Trpβ37(C3)

Trpβ37(C3) is located at the $\alpha_1\beta_2/\alpha_2\beta_1$ interfaces, at the contact between helix C from the β_2 (β_1) subunit and segment FG from the α_1 (α_2) subunit. In the T $\rightarrow$ R quaternary structure change, this contact is identified as the hinge (or flexible joint) region. This is because, in this region, the relative motions of the two $\alpha\beta$ dimers are small (<2 Å), and the packing between dimers does not change significantly.[36] Nevertheless, the side chains of several amino acid residues in this region, such as Trpβ37, change their conformation. In the deoxy-T structure, the indole $N_{\varepsilon1}H$ of Trpβ37 forms an intersubunit hydrogen bond with Aspα94(G1). In the ligated-R structure, this hydrogen bond is broken and, instead, the $N_{\varepsilon1}H$ group of Trpβ37 forms a hydrogen bond with Asnβ102(G4) within the same

[36] J. Baldwin and C. Chothia, *J. Mol. Biol.* **129**, 175 (1979).
[37] C. Ho, *Adv. Protein Chem.* **43**, 153 (1992).
[38] G. K. Ackers, *Adv. Protein Chem.* **51**, 185 (1998).
[39] M. R. Mihailescu and I. M. Russu, *Proc. Natl. Acad. Sci. USA* **98**, 3773 (2001).
[40] M. R. Mihailescu, C. Fronticelli, and I. M. Russu, *Proteins Struct. Funct. Genet.* **44**, 73 (2001).

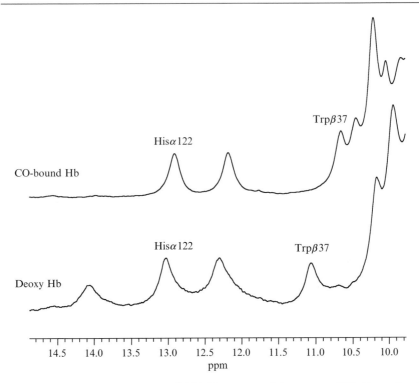

FIG. 5. Expanded region of the 1D ^{1}H NMR spectrum of deoxygenated Hb *(bottom)* and CO-bound Hb *(top)*, showing exchangeable proton resonances. Resonances of Hisα122 N$_{\varepsilon 2}$H and Trpβ37 N$_{\varepsilon 1}$H are indicated.

β subunit. The resonance of N$_{\varepsilon 1}$H of Trpβ37 can be resolved in the ^{1}H NMR spectrum downfield from the envelope of backbone NH resonances[40,41] (Fig. 5). The exchange of this hydrogen is slow enough to be observable in H–D exchange NMR measurements. Representative examples of the exchange curves are shown in Fig. 6. It is readily seen that in deoxygenated Hb the exchange takes $\sim$100 h [corresponding exchange rate, $(9.0 \pm 0.4) \times 10^{-6}$ s^{-1}] whereas in ligated Hb the exchange is complete in less than 250 min [corresponding exchange rate, $(3.3 \pm 0.2) \times 10^{-4}$ s^{-1}]. Therefore, the exchange of the indole hydrogen in Trpβ37 is sensitive to the allosteric transition from the deoxy-T to the ligated-R structure. Under the conditions of the experiment (namely, 0.1 M phosphate buffer with 0.1 M NaCl at pH 6.75), the exchange of this hydrogen

[41] V. Simplaceanu, J. A. Lukin, T.-Y. Fang, M. Zou, N. T. Ho, and C. Ho, *Biophys. J.* **79**, 1146 (2000).

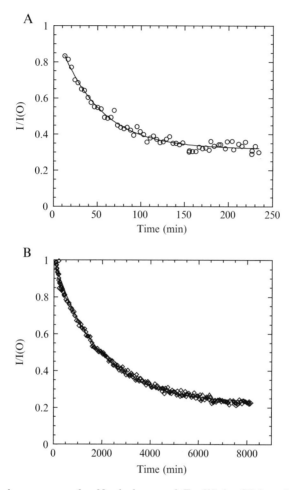

FIG. 6. Exchange curves for $N_{\varepsilon 1}$ hydrogen of Trpβ37 in CO-bound Hb (A) and deoxygenated Hb (B) in 0.1 M phosphate buffer containing 0.1 M NaCl at pH 6.75 and at 15°. The exchange was measured in H–D exchange experiments and the exchange rates are $(3.3 \pm 0.2) \times 10^{-4}$ s^{-1} in CO-bound Hb and $(9.0 \pm 0.4) \times 10^{-6}$ s^{-1} in deoxygenated Hb.

is in the EX2 regime.[40] Then, as for the DNA triple helices discussed in the previous section, one can calculate the change in stabilization free energy at the Trpβ37 sites on ligand binding to Hb directly from the exchange rates [Eq. (13)] as

$$\delta\Delta G_{\text{op}} = \Delta G_{\text{op}}(\text{deoxy}) - \Delta G_{\text{op}}(\text{ligated}) = 4.1 \text{ kcal/mol}$$

This result shows that, at each Trpβ37 site (at the $\alpha_1\beta_2$ and at the $\alpha_2\beta_1$ interface), the deoxy-T structure is energetically favored by $\sim$2 kcal/mol relative to the ligated-R structure.

The exchange of the indole hydrogen of Trpβ37 in Hb presents an interesting question regarding the mechanism of exchange in multisubunit proteins. The question arises due to the location of this residue at the interface between the two $\alpha\beta$ dimers. Accordingly, one could envision that the initial step in the opening reaction is the dissociation of the Hb tetramer into dimers. This mechanism would predict that the equilibrium constant of the opening reaction (K_{op}) for this hydrogen should be comparable to, or lower than, the equilibrium constant for tetramer–dimer dissociation ($K_d^{\alpha\beta}$). A simple experimental test shows that this is not the case. The tetramer–dimer dissociation constant for CO-bound Hb is 1.3×10^{-6} M (in 0.1 M Tris buffer containing 0.1 M NaCl and 1 mM EDTA, at pH 7.4 and at 15°). Under the same experimental conditions, the exchange rate of the $N_{\varepsilon 1}$ hydrogen of Trpβ37 in Hb is 1.2×10^{-3} s^{-1} and that in free Trp is 11 s^{-1}. These exchange rates predict [Eq. (2)] an equilibrium constant for opening of 1.1×10^{-4}. This value is two orders of magnitude higher than $K_d^{\alpha\beta}$. Therefore, the exchange of the Trpβ37 hydrogen probes an opening reaction that is distinct, and more favorable energetically, than the dissociation of the tetramer.

Hisα122(H5)

Hisα122(H5) is located at the intradimeric $\alpha_1\beta_1/\alpha_2\beta_2$ interfaces, where its side-chain $N_{\varepsilon 2}H$ group forms a water-mediated hydrogen bond with the side chain of Tyrβ35(C1) from the neighboring β subunit.[36] The NMR resonance of this hydrogen-bonded $N_{\varepsilon 2}H$ proton is well resolved in the 1D NMR spectrum[41] (Fig. 5). Consistent with the structural invariance of the $\alpha_1\beta_1/\alpha_2\beta_2$ interfaces on ligand binding, the spectral position of this resonance in deoxygenated Hb is the same as that in ligated Hb. The exchange of this hydrogen occurs on the millisecond time scale and, thus, can be measured in transfer of magnetization experiments. The dependence of the exchange rate on pH is shown in Fig. 7. The exchange rate increases with increasing pH, indicating that the exchange is OH$^-$ catalyzed. A plot of the same exchange rate data as a function of the concentration of OH$^-$ (Fig. 7B) has the hyperbolic form predicted by Eqs. (1) and (4). The algebraic form of Eq. (1) is such that not all its parameters can be uniquely determined from the nonlinear least-squares fit of the exchange rate as a function of catalyst (OH$^-$) concentration. For this reason, in Fig. 7, we have chosen to express some of these parameters relative to the rate constant for OH$^-$-catalyzed exchange (k_{OH}). The fit shown in Fig. 7 also

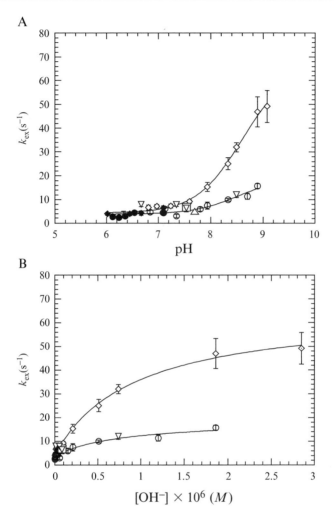

FIG. 7. Exchange rates of $N_{\varepsilon 2}$ hydrogen of Hisα122 obtained from transfer of magnetization experiments in 0.1 M Bis-Tris buffer (solid symbols) or 0.1 M Tris buffer (open symbols) with 0.18 M NaCl at 37°. (A) Dependence on pH. (B) Dependence on concentration of OH^-. (♦, ◇) deoxygenated Hb; (●, ○) CO-bound Hb; (▼, ▽) O_2-bound Hb; (□) CN-met Hb; (△) azido-met Hb. The curves were obtained by fitting the exchange rate as a function of the concentration of OH^- to Eq. (1), in which $k_{ex,open}$ was expressed as in Eq. (4) with $k_H = 0$ and $k_B^{(i)} = 0$. The fitted parameters are as follows: $k_{op} = 63 \pm 2$ s^{-1}, $k_w/k_{OH} = (5.7 \pm 0.8) \times 10^{-7}$ M, and $(k_{op} + k_{cl})/k_{OH} = (7.2 \pm 0.8) \times 10^{-6}$ M for deoxygenated Hb and $k_{op} = 19 \pm 3$ s^{-1}, $k_w/k_{OH} = (1.6 \pm 0.6) \times 10^{-6}$ M, and $(k_{op} + k_{cl})/k_{OH} = (6.6 \pm 4.5) \times 10^{-6}$ M for ligated Hb.

assumes that the structural opening reaction involved in the exchange of Hisα122 $N_{\varepsilon2}H$ is not affected by increasing concentrations of catalyst (i.e., increasing pH). This assumption cannot be made *a priori*, and its validity should be checked experimentally to the best extent possible. For example, for Hisα122 in Hb, the position of its $N_{\varepsilon2}H$ resonance is constant as a function of pH. Moreover, the R_1 rate of the Hisα122-$N_{\varepsilon2}$ proton and the NOEs from this proton to neighboring protons are also pH independent.[39] These results support the assumption that increasing pH does not induce significant conformational changes at the Hisα122 site.

The interesting feature of the results shown in Fig. 7 is that the exchange rate of the Hisα122 $N_{\varepsilon2}$ hydrogen in deoxygenated Hb is higher than in ligated Hb. This result implies that the Hisα122 site is sensitive to the T → R allosteric transition. The allosteric change at this site affects the opening rate of Hisα122, namely, the opening rate k_{op} decreases from 63 s^{-1} in deoxygenated Hb to 19 s^{-1} in ligated Hb. This change can be recast in terms of a change in structural stabilization free energy based on Eq. (14) as

$$\delta\Delta G_{op} = \Delta G_{op}(\text{deoxy}) - \Delta G_{op}(\text{ligated}) = -1.4 \text{ kcal/mol}$$

One notes that the sign of this change is the opposite of that at the Trpβ37 sites: on the transition from the deoxy-T to the ligated-R state, the stabilization free energy at the Hisα122 sites increases whereas that at the Trpβ37 sites decreases.

The hydrogen exchange results for Hisα122 in Hb reiterate the complementarities between structure- and site-specific energetics noted earlier for DNA triple helices. Hisα122 belongs to a part of the Hb structure that is structurally invariant in ligand binding. Nevertheless, the exchange rate of the Hisα122 hydrogen changes on binding of ligand, suggesting that the T → R allosteric transition affects the energetics at the Hisα122 sites. The results also reveal the complex pattern of changes in hydrogen exchange that can be induced in a protein by a functional transition. As detailed above, in Hb, hydrogens such as peptide NH and Trpβ37 $N_{\varepsilon1}H$ sense the transition from T (tense) to R (relaxed) thermodynamic states by an acceleration of their exchange rates. At other sites, such as Hisα122, 122, the same T → R global transition slows down hydrogen exchange, even in the absence of a change in local structure.

Summary

The results presented in this chapter illustrate the use of hydrogen exchange and NMR spectroscopy to define how the free energy of stabilization of a biological molecule is being redistributed or lost as a result of a structural perturbation or of a functional change. The methodology relies on the potential of hydrogen exchange measurements to delineate and to quantify structural energetic changes throughout the molecule. At the same time, the methodology relies also on the resolving power of proton NMR spectroscopy. This resolving power generally decreases with increasing the size of the molecule observed. In large molecules, special environments of the proton can result in large shifts of its resonance. In these cases, observation of the proton by 1D NMR is possible, like in the case of hemoglobin discussed here. In general, however, at the present time, NMR spectroscopy can be used to monitor hydrogen exchange in molecules and molecular complexes of moderate size. Advances in isotopic labeling of proteins and nucleic acids, and in NMR techniques, promise to extend this limit to larger macromolecular systems. It is certain that these developments will provide the basis for new illuminating analyses of site-specific structural energetics by hydrogen exchange and NMR spectroscopy.

Acknowledgments

The work presented in this chapter has been supported by the National Science Foundation (MCB-9723694) and by the American Heart Association.

[10] Fluorescence Quenching Methods to Study Protein–Nucleic Acid Interactions

By SIDDHARTHA ROY

Introduction

The study of protein–nucleic acid interaction is at the heart of many biological phenomena. Dissociation constants of protein–nucleic acid complexes range from 10^{-12} M for lac repressor–operator interaction[1] to 10^{-4} M for some tRNA–synthetase interactions[2] and stoichiometry ranges from many proteins binding to a single ligand, as in the case of DNA-binding proteins[3] to several ligands binding to a single protein, as in the case tRNA binding to proteins.[4] Clearly no single technique or strategy is

[1] D. E. Frank, R. M. Saecker, J. P. Bond, M. W. Capp, O. V. Tsodikov, S. E. Melcher, M. M. Levandoski, and M. T. Record, Jr., *J. Mol. Biol.* **267,** 1168 (1997).
[2] S. S. Lam and P. R. Schimmel, *Biochemistry* **14,** 2775 (1975).
[3] D. F. Senear, M. Brenowitz, M. A. Shea, and G. K. Ackers, *Biochemistry* **25,** 7344 (1986).
[4] I. Gruic-Sovulj, H. C. Ludemann, F. Hillenkamp, I. Weygand-Durašević, Z. Kucan, and J. Peter-Katalinic, *J. Biol. Chem.* **272,** 32084 (1997).

suitable for studying such diversity. Many methods have been devised, such as quantitative footprinting, optical spectroscopy, nuclear magnetic resonance (NMR), electrophoresis, and ultracentrifugation, to name a few. Each technique has its own advantages and disadvantages and the range of dissociation constant in which it works best. Fluorescence spectroscopy has been widely used to study ligand–protein interaction because of its simplicity and sensitivity. Fluorescence methods are particularly popular because they are well suited for protein–ligand interactions with dissociation constants in the range of 10^{-4} to 10^{-8} M, where some protein–nucleic acid dissociation constants fall.

Except in specialized cases such as Y base in yeast tRNA[Phe], nucleic acids do not possess any intrinsic fluorescence. In these situations, binding can be studied by quenching of tryptophan or tyrosine fluorescence of the proteins. Most proteins contain one or more tryptophan residues, although many are known to be devoid of tryptophan residues. In these cases fluorescence from tyrosine residues can be utilized to study ligand binding. Only a small number of known proteins contain neither tryptophan nor tyrosine, such as bacterial histone-like protein HU. Ligand binding to this latter class of proteins cannot be studied by the methods described here. We must emphasize that it is difficult to measure the intrinsic fluorescence of proteins reliably at a concentration lower than 10^{-8} M. Hence, it is difficult to measure dissociation constants in the subnanomolar range, where many operator–repressor interactions fall.[1]

Many fluorescence parameters, such as anisotropy, intensity, and energy transfer efficiency, are sensitive to formation of protein–ligand complexes and can be utilized to derive binding isotherms.[5] The change in fluorescence intensity on formation of a protein–ligand complex is one of the simplest ways to measure binding of a ligand to a protein. It offers a way to estimate the interaction free energy at equilibrium without any special assumptions. Such equilibrium measurements are vital for obtaining thermodynamic parameters. In addition to obtaining equilibrium constants, fluorescence quenching methods can provide clues to conformation of the protein–nucleic acid complexes. External collisional quenchers offer another way to detect conformational differences of protein–nucleic acid complexes. We divide the chapter into two parts: the first deals with issues related to the measurement of binding affinity and the second deals with the elucidation of conformation of the nucleic acid-bound proteins.

[5] L. D. Ward, *Methods Enzymol.* **117,** 400 (1985).

Materials

Quartz Curvettes

Standard 1 × 1 cm fluorescence cuvettes may be used for quenching studies, although this will require higher volumes and considerably more materials; 0.5 × 0.5 cm fluorescence cuvettes may be used, and require approximately one-fourth the material of the standard cuvette. This has an additional advantage that, because of the shorter path length, the inner filter effect is reduced significantly. Even smaller volume cuvettes may be used. The author has successfully used a cuvette, requiring only 45 μl, in a PerkinElmer Life and Analytical Sciences (Boston, MA) spectrofluorometer. These types of cuvettes are now available from many commercial sources.

Solutions

Most commonly used buffers such as phosphate or Tris are compatible with fluorescence quenching studies. For others, it is important to know whether the buffer is transparent in the 295- to 340-nm range. High absorbance of buffer in this region is generally incompatible with quenching studies. It is always desirable to check the fluorescence spectrum of the buffer alone to make sure no significant background fluorescence is present. It is preferable to have protein solution dialyzed and the ligand dissolved in the same buffer. Even though the ligand is not expected to have fluorescence, it is possible and sometimes observed that ligand solutions have weak fluorescence from contaminating impurities. It is always wise to carry out a blank titration without the protein. If fluorescence from ligand is observed, appropriate subtraction should be carried out.

Estimation of Binding Affinity

Monitoring Protein Fluorescence Titration with Ligand

Tryptophans can be excited only in proteins that contain tryptophan. Excitation at 295 nm results in selective excitation of tryptophan residues. For increased sensitivity, excitation at 280 nm can be used, although it results in excitation of other aromatic amino acid residues such as tyrosine. More importantly, nucleic acids have much higher absorbance at 280 nm and the inner filter effect is a much more serious problem at this wavelength. It is desirable to record full emission spectra and the emission maximum after addition of the ligand, although the measurement of fluorescence at a single wavelength is adequate. In case the tryptophans are photosensitive, the light exposure can be minimized by using the smallest slit opening on the excitation side and enlarging the emission slit to compensate for the loss of sensitivity.[6] In between readings, the excitation shutter can be closed to minimize exposure of the sample to light. Either fluorescence intensity at a single emission wavelength or total integrated emission intensity can be used as a measure of fluorescence quenching. Titrations carried out by an addition of ligand causes dilution of the protein and hence reduction of fluorescence intensity. It is always desirable to keep this volume dilution to a minimum. Under such circumstances, a simple correction factor based on volume may be used for obtaining a correct measure of fluorescence quenching.

An important question concerns how long one should wait after addition of the ligand before carrying out a fluorescence measurement. Most ligands have high association rate constants and generally equilibrate rapidly. A wait of 1–2 min is generally sufficient to equilibrate for most systems. However, slow binding kinetics of even small ligands are known, the two well-known ones being colchicine binding to tubulin[7] and bis-(8-aniline-1-naphthalenesulfonate) binding to proteins.[8] The half-life of association in the latter case is temperature dependent and at lower temperatures may reach hours. Thus, in general, it is prudent to check whether there is any time dependence of fluorescence intensity on mixing of the protein and the ligand.

Effect of Protein Concentration on Measurement of Dissociation Constant

The choice of protein concentration in the determination of binding constants by quenching of internal fluorescence needs further elaboration. In all ligand-binding experiments, protein concentration is held constant and increasing concentrations of ligand are added. If the total protein concentration is much less than the dissociation constant of the protein–ligand complex, then the free ligand concentration $[L_f] \gg \sum i \cdot [P \cdot L_i]$, where $[P \cdot L_i]$ values are the concentrations of different protein–ligand complexes. In such a case the quenching profile is devoid of any significant stoichiometric information and the dissociation constant can be extracted directly. When the protein concentration is much higher than the dissociation constant of the protein–ligand complex, virtually all added ligand will form complex until saturation. In this case, only stoichiometric information is

[6] S. Chatterjee, Y. N. Zhou, S. Roy, and S. Adhya, *Proc. Natl. Acad. Sci. USA* **94,** 2957 (1997).
[7] B. Bhattacharyya and J. Wolff, *Proc. Natl. Acad. Sci. USA* **71,** 2627 (1974).
[8] A. Bhattacharyya, A. K. Mandal, R. Banerjee, and S. Roy, *Biophys. Chem.* **87,** 201 (2000).

present at the break point, which is devoid of any information about the binding constant. These situations are shown in Fig. 1. Clearly it is difficult to extract any information about the dissociation constant with protein concentrations about an order of magnitude higher than the dissociation constant. If the protein concentration is in between, information about both stoichiometry and the binding constant is present and can be extracted.

Thus, the choice of protein concentration is critical for estimating the dissociation constant. It is preferable to have the protein concentration below the dissociation constant of the protein–ligand complex. In systems where ligand–protein dissociation constants are so low that one is compelled to work at submicromolar concentrations, the stability of the protein, both biological and spectroscopic, becomes important. In several cases in which the protein concentration was kept low, we have seen red shifting of emission maxima on incubation at temperatures above $30°$. This may be a consequence of lower stability of some proteins at low concentrations and may be catalyzed by solid surface. The stability of the protein

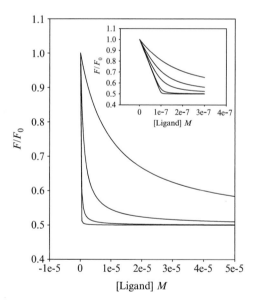

FIG. 1. Simulation of the effect of relative values of protein concentration and the dissociation constant. The protein concentration was fixed at $10^{-7}M$ and the curves from right to left are for K_d values of 10^{-5}, 10^{-6}, 10^{-7}, and $10^{-8}M$, respectively. *Inset:* The stoichiometric points in detail. The curves are for K_d values of 10^{-7}, 3×10^{-8}, 10^{-8}, 10^{-9}, and $10^{-10}M$ from right to left. The simulation was carried out with SigmaPlot, using a single-site binding equation.

must be ascertained before proceeding with the titration, particularly if the titration is carried out at low protein concentrations and relatively high temperatures. The sensitivity of currently available spectrofluorometers allows use of protein concentration up to about 10^{-8} M, depending on the tryptophan content. Hence ligand protein dissociation constants in the subnanomolar range are at present inaccessible by this method.

Inner Filter Effect Correction. The biggest problem in the measurement of internal aromatic fluorescence quenching comes from the inner filter effect. Many ligands have strong absorbance in the range of excitation and emission wavelengths normally used for tryptophan fluorescence measurements (280–295 and 330–350 nm, respectively). The inner filter effect arises from the fact that although a typical cuvette has a cross-section of 10 × 10 mm, the actual volume from which the fluorescence is observed is in the middle of the cross-section and is small. Before the exciting light reaches that volume or the emitted light reaches the photomultiplier tube, the light must travel through the absorbing solution. If the absorbance is high, the light intensity reaching the active volume or the photomultiplier tube is reduced. Thus the absorbing solution acts as a filter and the effect is known as the inner filter effect. If the ligand has absorbance at the emission or excitation wavelength, it is mandatory that a correction for the inner filter effect be employed. The simplest correction for inner filter effect is given by the following formula and is widely used.

$$F_{corr} = F_{obs} \cdot 10^{(A_{ex}+A_{em}) \cdot L/2} \tag{1}$$

where L is the pathlength of the cuvette used, the A terms are the absorbances at the excitation and emission wavelengths, and the F values are the corrected and observed fluorescence intensities. Equation (1) is derived by assuming that light from only an infinitesimal volume at the center of the cell in a 10 × 10 mm cell reaches the photomultiplier tube. In reality, however, light from a significant volume element reaches the photomultiplier tube. Hence, this correction formula should be used with caution and should be considered valid for relatively low absorbance values only. It is generally advisable to limit absorbance values to less than 0.3. If the use of high absorbance becomes unavoidable, shorter pathlength cuvettes (e.g., cross section 0.5 × 0.5 cm) should be employed.

Data Analysis

The most straightforward way of analyzing data is to use a nonlinear least-squares fit procedure. Three basic assumptions must be fulfilled in order to use the nonlinear least-squares fit procedure: (1) the source of major experimental errors of the data is confined to the vertical axis, that is, the determination of fluorescence intensity; (2) the errors follow normal distribution; and (3) there is no systematic error in the data. If conducted properly all three assumptions are fulfilled in fluorescence quenching experiments and hence nonlinear least-squares fit can be used to extract binding constants. The nonlinear least-squares fit procedure for analyzing binding data has been discussed in detail.[9] Many of the commercially available fitting programs have a nonlinear least-squares fit program as part of the package. The binding data can be fitted to Eq. (2) (assuming F_{obs}/F_0 is being measured and the starting ratio is 1; see the next section):

$$F_{ratio} = 1 - \Phi \left[\frac{(K_d + nP_0 + L_T) - \sqrt{(K_d + nP_0 + L_T)^2 - (4nP_0 \cdot L_T)}}{2nP_0} \right] \quad (2)$$

where F_{ratio} is the observed fluorescence ratio, Φ is the fluorescence ratio change amplitude ($= 1 - F_{ratio}^{\infty}$), P_0 is the protein concentration, K_d is the dissociation constant, n is the stoichiometry, and L_T is the total ligand concentration. As can be seen in Eq. (2), there are three unknown parameters: K_d, n, and Φ. The preferable procedure then involves treating n, Φ and K_d as the fitting parameters so that an accurate determination of Φ is not necessary. We generally prefer to determine stoichiometry also by procedures described elsewhere[10] and treat K_d and Φ only as fitting parameters.

Example: tRNA^{Gln} Binding to Glutaminyl-tRNA Synthetase

Glutaminyl-tRNA synthetase (GlnRS) binds one molecule of cognate tRNA^{Gln}. Fluorescence quenching offers one of the best equilibrium methods to study the binding and variation of binding constants under various solution conditions.

tRNA: tRNA^{Gln} is purified from an overproducing strain, using column chromatography as described.[10] The tRNA^{Gln} has a specific activity higher than 1.6 nmol/A_{260} unit. The tRNA^{Gln} is dialyzed against water and lyophilized. The lyophilized tRNA is dissolved in appropriate buffer before titration

Protein: Glutaminyl-tRNA synthetase is purified from an overproducing strain as described.[10] The enzyme is dialyzed against 0.1 M Tris-HCl, pH 7.5, for 18 h at 4° before use

[9] M. L. Johnson and S. G. Frasier, *Methods Enzymol.* **117**, 301 (1985).
[10] T. Bhattacharyya, A. Bhattacharyya, and S. Roy, *Eur. J. Biochem.* **200**, 739 (1991).

Spectrofluorometer: A single-channel fluorometer (F-3010; Hitachi, Tokyo, Japan) is used. It is generally a good precaution to check the stability of the fluorometer over a reasonable time period on Raman scattering band of water. Any major drift would significantly compromise data quality. A bandpass of 5 nm is used for excitation and emission

Method. The excitation wavelength is set at 295 nm. The protein and the tRNA are diluted from the concentrated stock to reduce pipetting error. Often a $3\times$ stock of protein and a $10\times$ stock of tRNA are used (3 and 10 times the final desired concentrations). The initial concentration of GlnRS is kept at 500 nM. In general, we have observed that the GlnRS concentrations below this level give a time-dependent red shift of unknown origin. It could be due to surface-catalyzed denaturation, which has been described in studies of other proteins. The protein is first diluted in the cuvette which has been incubated for some time to equilibrate the temperature. A fluorescence emission spectrum is then recorded from 300 to 400 nm. This is followed by addition of the required volume of tRNA solution, which is mixed carefully, and the emission spectrum is observed. The whole process is then repeated with new protein and tRNA solutions at different concentrations. A blank buffer reading is taken at the start of the experiment and at the end and subtracted from the fluorescence value.

The fluorescence ratios at 340 nm obtained by this method are plotted as a function of tRNA concentration and fitted to Eq. (2) by a procedure described previously. Figure 2 shows two typical quenching curves obtained by the above-described procedure.

Elucidation of Conformation of Protein–Nucleic Acid Complexes

Many DNA-binding proteins are known that form complexes with different target DNA sequences, that is, they are capable of recognizing different DNA sequences. How a protein recognizes multiple sequences is not known. It has been suggested that at least some DNA-binding proteins are flexible, so that they can model themselves on different target sequences.[11,12] Fluorescence quenching can be a powerful tool to differentiate between different types of complexes on the basis of quantum yield or other fluorescence parameters. In multitryptophan proteins, complex formation with different target sequences may lead to a change in tryptophan fluorescence intensity, which may be compensating in different complexes. Collisional quenching is capable of resolving different classes of

[11] R. Reeves and L. Beckerbauer, *Biochim. Biophys. Acta* **1519,** 13 (2001).
[12] T. K. Kerppola, *Structure* **6,** 549 (1998).

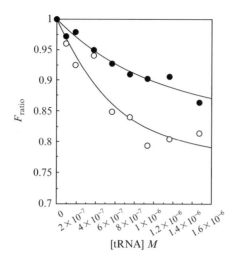

Fig. 2. Plot of F_{ratio} versus tRNA^{Gln} concentration GlnRS (0.5 μM) was titrated with increasing concentrations of tRNA^{Gln} and the fluorescence value at each point was determined. The experiments were conducted in 0.01 M Tris-HCl buffer, pH 7.5, containing 50 mM (open circles) and 100 mM (solid circles) KCl. The solid lines are best-fit lines. The temperature was 25°. The excitation wavelength was 295 nm and emission was recorded at 340 nm.

tryptophans on the basis of their accessibilities and thus provides significantly greater power to resolve differences in quantum yield and emission maximum. Even more importantly, differences in accessibility (as characterized by Stern–Volmer constants or bimolecular quenching constant, k_q) in different complexes provide another parameter by which conformational differences may be identified.[13]

Principles of Quenching Method and Data Processing

Most collisional quenching studies are carried out by observing the fluorescence intensity and emission maximum at increasing quencher concentrations. Because of the sensitivity of many protein–nucleic acid complexes to different ions, it is difficult to use iodide or other ions for quenching studies. It is well known that many protein–nucleic acid complex stabilities are a function of the nature of the anion or cation.[14] Thus, an

[13] M. R. Eftink and C. A. Ghiron, *Anal. Biochem.* **114,** 199 (1981).
[14] J. H. Ha, M. W. Capp, M. D. Hohenwalter, M. Baskerville, and M. T. Record, *J. Mol. Biol.* **228,** 252 (1992).

equivalent concentration of another salt may not be an appropriate control. For this reason, the author prefers to use neutral quenchers such as acrylamide.

The quenching studies are usually carried out in a 1 × 1 cm cuvette. If high concentrations of acrylamide are desired, it is better to use a shorter pathlength cuvette because of the inner filter effect (see below). An excitation wavelength of 295 nm is often used. At this wavelength, tryptophans are primarily excited without too much loss in sensitivity (the molar extinction coefficient of tryptophan is approximately 50% at 295 nm compared with 280 nm). It is preferable to record the whole emission spectrum (typically from 300 to 450 nm) at each concentration of acrylamide and determine the emission maximum. In multitryptophan proteins, shift of emission maximum can sometimes give valuable information regarding various tryptophans and their environment.

Acrylamide has significant absorbance at 295 and 280 nm. The inner filter effect is, thus, a serious problem for acrylamide quenching studies. Because of the complex nature of the inner filter effect correction at higher absorbances, it is probably preferable to keep acrylamide concentrations below 200 mM or to use a shorter pathlength cuvette. Use of lower acrylamide concentrations may also address the possible effect of higher concentration of acrylamide on the dissociation constant of the complex.

For single-tryptophan proteins, the quenching data can be plotted as F_0/F versus [quencher]. The slope of the line gives the Stern–Volmer constant, K_{SV}.

$$F_0/F = 1 + K_{SV}[Q] \tag{3}$$

In multitryptophan proteins, the situation is more complex. The quenching relationship is described by Eq. (4):

$$F_0/F = \{f_1/(1 + K_{SV}^1[Q]) + f_2(1 + K_{SV}^2[Q])\}^{-1} \tag{4}$$

Where f_1 is the fraction of fluorescence intensity contributed by the first tryptophan, f_2 is the fraction of fluorescence intensity contributed by the second tryptophan, and the superscript indices on the Stern–Volmer constants reflect the Stern–Volmer constants of the respective tryptophans.

There are a number of ways by which information about quenchabilities of different tryptophans can be obtained. A useful modification of the Stern–Volmer equation in the case of multitryptophan proteins, where one tryptophan is quenchable and the others are not, is the Lehrer equation:

$$F_0/\Delta F = 1/f_1 K_{SV}^1[Q] + 1/f_1 \tag{5}$$

where f_1 is the fraction of the total fluorescence that is quenchable, ΔF is $F_0 - F$, and K_{SV}^1 is the Stern–Volmer constant of the quenchable tryptophan. A plot of $F_0/\Delta F$ versus $1/[Q]$ yields an intercept of $1/f_1$ and a slope of $1/f_1 K_{SV}^1$, from which the Stern–Volmer constant can be calculated.[15]

A better way to analyze the quenching data is to fit the data to equations with an increasing number of quenchable fluorophores until a satisfactory fit is obtained. Many currently available commercial programs, such as SigmaPlot (SPSS, Chicago, IL) or KyPlot (KyensLab, Tokyo, Japan) can be used for such fitting. In our experience, a good guess concerning the values of the starting parameter is important for obtaining a correct fit.

Example: Complexes of λ Repressor with Various Operator Sites

λ repressor binds to six operator sites, O_R1, O_R2, O_R3, O_L1, O_L2, and O_L3, within the bacteriophage λ genome with different binding constants. The aim is to find out whether the conformations of the repressor bound to O_R1, O_R2, and O_R3 are different. All three tryptophans are situated in the C-terminal domain, distant from the N-terminal DNA-binding domain. Acrylamide quenching of tryptophan fluorescence is used to elucidate the conformational differences of repressor bound to three different operator sites.

Oligonucleotides

```
O_R1:5' - AAC CAT TAT CAC CGC CAG AGG TAA AAT AG-3'
          TTG GTA ATA GTG GCG GTC TCC ATT TTA TC

O_R2:5' - AAC CAT TAA CAC CGT GCG TGT TGA AAT AG-3'
          TTG GTA ATT GTG GCA CGC ACA ACT TTA TC

O_R3:5' - AAC CAT TAT CAC CGC AAG GGA TAA AAT AG-3'
          TTG GTA ATA GTG GCG TTC CCT ATT TTA TC
```

The oligonucleotides are purified by reversed phase HPLC on a Waters UniFord, MA C_{18} μBondapak column using 0.1 M triethylammonium acetate, pH 7, as mobile phase and an acetonitrile gradient. Extinction coefficients are calculated by using 14,000 for purines and 7000 for pyrimidines.[16] The complementary oligonucleotides are mixed at a molar ratio of 1:1 in 0.1 M potassium phosphate buffer, pH 8.0, and annealed by heating to $80°$ followed by slow cooling

---

[15] S. S. Lehrer, *Biochemistry* **10**, 3254 (1971).
[16] S. Roy, S. Weinstein, B. Borah, J. Nickol, E. Appella, J. L. Sussman, M. Miller, H. Shindo, and J. S. Cohen, *Biochemistry* **25**, 7417 (1986).

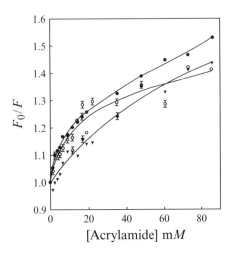

[Acrylamide] mM

FIG. 3. Stern–Volmer plot of acrylamide quenching of λ repressor bound to single-operator sites O_R1 (●), O_R2 (▼), and O_R3 (○). The solid lines are the best-fit lines to the Stern–Volmer equation [Eq. (4)]. The measurements were done with a 0.5 μM concentration of oligonucleotide and protein dimer in 0.1 M potassium phosphate buffer, pH 8.0. The observed fluorescence was corrected for the inner filter effect as described in text (see Inner Filter Effect Correction). The excitation wavelength was 295 nm, and emission was measured at 340 nm. The experiments were repeated three or more times. Reproduced from S. Deb, S. Bandyopadhyay, and S. Roy, *Biochemistry* **39,** 3377 (2000) (with permission from the American Chemical Society).

Protein: λ repressor is purified as described[17] and estimated to be greater than 95% pure. The protein is dialyzed against 0.1 M potassium phosphate buffer, pH 8.0, for 18 h at 4°. The concentration of the protein is determined, using an $E_{280}^{1\%}$ value of 11.8

Acrylamide: Stock acrylamide solution is prepared by dissolving an accurately weighed amount of three times-recrystallized solid acrylamide in 0.1 M potassium phosphate buffer, pH 8.0, so that the desired final solution volume is attained. The desired stock concentration is made to minimize protein dilution while maintaining the accuracy of the volume to be pipetted. Depending on the situation, 1 to 5 M stock solutions are used

Spectrofluorometer: A single-channel fluorometer (Hitachi F-3010) is used. It is generally a good precaution to check the stability of the fluorometer over a reasonable time period on Raman scattering

[17] R. Saha, U. Banik, S. Bandyopadhyay, N. C. Mandal, B. Bhattacharyya, and S. Roy, *J. Biol. Chem.* **267,** 5862 (1992).

TABLE I
STERN–VOLMER CONSTANTS OF λ REPRESSOR BOUND TO
VARIOUS SINGLE OPERATOR SITES

	K_{sv}^1 (M^{-1})	f_1	K_{sv}^2 (M^{-1})	f_2
O_R1	214 ± 15	0.2 ± 0.015	2.51 ± 0.26	0.78 ± 0.017
O_R2	31.5 ± 10.1	0.48 ± 0.1	0.35 ± 0.3	0.52 ± 0.1
O_R3	129 ± 71	0.25 ± 0.04	1.01 ± 0.31	0.75 ± 0.04

band of water. Any major drift would significantly compromise data quality

Method. The excitation and emission wavelengths are set at 295 and 340 nm, respectively. A bandpass of 5 nm is set for both excitation and emission. The annealed duplex oligonucleotide containing the desired operator site is mixed with protein in 0.1 M potassium phosphate, pH 8.0, so that the final concentration of both the oligonucleotide and the protein (in terms of dimer) is 0.5 μM. This concentration of protein is far higher than the dissociation constant of the dimer–operator complex while significantly lower than the tetramer–dimer dissociation constant. Acrylamide quenching of tryptophan fluorescence is done by adding a freshly prepared high concentration of three times-recrystallized acrylamide in 0.1 M potassium phosphate buffer, pH 8.0, to the protein solution. Appropriate blank spectra are subtracted from the actual spectra. If acrylamide solution has weak fluorescence, corresponding control spectra without protein should be subtracted. Otherwise, subtraction of buffer blank is adequate. All fluorescence values are corrected for volume changes and the inner filter effect, using Eq. (1). The quenching curves and the extracted K_{SV} values are shown in Fig. 3 and Table I.

[11] Thermodynamics, Protein Modification, and Molecular Dynamics in Characterizing Lactose Repressor Protein: Strategies for Complex Analyses of Protein Structure–Function

By LISKIN SWINT-KRUSE and KATHLEEN S. MATTHEWS

Introduction

Equilibrium measurements of protein function often falsely evoke a perception of molecular stasis—with the protein limited to the beginning and ending states of a reaction or providing a static scaffold to support specific activity. In fact, thermodynamically observed allostery and cooperativity provided one of the first indications that protein structure plays an active role in executing function. Several previous chapters in this series provide detailed information about analyzing thermodynamics of protein–DNA and protein–ligand interactions.[1–6] The information presented in this volume, and specifically in this chapter, focuses on strategies for understanding dynamic protein intermediates during allosteric and cooperative ligand binding.

Identification of intermediate species is crucial to a mechanistic understanding of allosteric proteins. For occupancy of one binding site to affect binding energetics at a second, distant site, the protein molecule must have an active, "in-house" system for communication that is transmitted through changing protein structure. However, identifying the components of this process is complicated by the potential for multiple types of intermediates, and even multiple behaviors for a single intermediate. These include (1) thermodynamically populated intermediates, which must be explicitly accounted for in analyses of energetic experiments, (2) partially liganded intermediates, which must exist as a protein with multiple binding sites approaches its fully liganded state, and (3) transient structural intermediates on the dynamic pathway between the initial and final protein states. Each type of intermediate may be detectable, depending on the modes and limits of detection. In lactose repressor protein (LacI), which

[1] I. Wong and T. M. Lohman, *Methods Enzymol.* **259,** 95 (1995).
[2] T. M. Lohman and D. P. Mascotti, *Methods Enzymol.* **212,** 400 (1992).
[3] T. M. Lohman and W. Bujalowski, *Methods Enzymol.* **208,** 258 (1991).
[4] M. T. Record, Jr., J.-H. Ha, and M. A. Fisher, *Methods Enzymol.* **208,** 291 (1991).
[5] L. Jen-Jacobson, *Methods Enzymol.* **259,** 305 (1995).
[6] E. Di Cera, *Methods Enzymol.* **232,** 655 (1994).

serves as the specific example in this chapter, partially liganded intermediates have eluded detection, and none of the intermediates appear to be highly populated. Nonetheless, as we present, significant information about the molecular changes that occur in the pathway between LacI states has been derived from integration of thermodynamic analysis with mutagenesis and molecular dynamics studies.

The ultimate description of an allosteric protein includes an atom-by-atom list of intermediates as the protein progresses from one state to another. These intermediates may vary with changing environmental conditions, therefore altering cooperative behaviors. To parse the allosteric network into contributions from its components—subunits, domains, side chains, and individual atoms—one must deconvolute the intertwined energetics. This process is particularly challenging for a protein for which the structure is not available. In some cases, information may be obtained for a part of the protein, but these data must be placed in the context of overall function. Ideally, structural data can be combined with careful experiments to assign contributions at the atomic level. Further, molecular dynamics simulations provide a means for interpolating the protein motions from structural snapshots.

Even within a family of proteins that share high sequence homology and engage in similar reactions, mechanistic details may vary considerably. An excellent example is the LacI family of genetic regulatory proteins—within this collection of bacterial genetic regulators are proteins that are inducible (e.g., LacI), repressible (e.g., PurR), or exert their influence primarily by interaction with other proteins (e.g., CytR).[7,8] This chapter uses studies of tetrameric lactose repressor protein (LacI) to illustrate strategies that combine thermodynamic analyses, mutagenesis, and molecular dynamics to decipher functional contributions from regions and specific sites. LacI has a long, rich history in the development of these strategies.[7] Now, with the coincident emergence of molecular dynamics and bioinformatics, LacI is poised to provide insight to these new frontiers in understanding the relationship between protein structure and functions.

LacI Function

LacI function can be most simply described as having two modes.[7] In the first, LacI regulates expression of the structural genes for the *lac* operon by associating with high affinity to its target operator sequence, O^1. To

[7] K. S. Matthews and J. C. Nichols, *Prog. Nucleic Acid Res. Mol. Biol.* **58**, 127 (1998).

[8] L. Søgaard-Andersen, N. E. Møllegaard, S. R. Douthwaite, and P. Valentin-Hansen, *Mol. Micriobiol.* **4**, 1595 (1990).

relieve repression, a small-molecule inducer binds a second site within each protein monomer to elicit a conformational change that diminishes affinity for O^1. Note that the DNA-binding site requires the dimeric unit, whereas each monomer contains an intact inducer-binding site. Thus, the LacI tetramer has two DNA-binding sites and four inducer-binding sites. The LacI functional cycle and structure are shown in Fig. 1.

Ligand Binding and Allostery: Detecting Functional Intermediates

The first steps in parsing thermodynamic contributions to different processes for a protein are to define end points, that is, the stoichiometries of fully liganded complexes (Scheme I). All functional activities must be assessed by direct thermodynamic measurements, including differentiating between specific and nonspecific ligand binding and protein association. Conditions for these binding experiments are crucial in interpreting the results, and one should be especially alert for indications of allosteric or cooperative binding events (see Scheme II). Using these types of approaches, significant insights into LacI structure/function and thermodynamics were possible well before the three-dimensional structures became available. Analysis of LacI thermodynamic states included experiments that measured operator DNA binding in the absence and presence of inducer and, conversely, inducer binding in the absence and presence of operator (Fig. 2).[9,10] Two LacI sites for operator DNA and four sites for inducer were deduced from binding studies under conditions to measure stoichiometry of interaction rather than affinity (see Fig. 2, inset).[11] As a consequence of stoichiometry measurements, LacI tetramer in complex with two DNA molecules was postulated to form looped DNA structures. This capacity has been demonstrated directly in multiple laboratories.[7]

Cooperativity in inducer binding and the reciprocal allosteric response of operator/inducer were readily demonstrated for LacI, as illustrated in Fig. 2.[9,10] Detection of cooperativity and allostery indicates that intra- and intersubunit communication must occur within the protein. Therefore, a physical pathway between the relevant binding sites must exist and, by definition, is populated by allosteric intermediates. For example, binding to inducer is cooperative for LacI in the presence of DNA at neutral pH.[10] Thus, a physical pathway must exist between the two separate inducer-binding sites of two monomers that comprise a dimer.

---

[9] R. B. O'Gorman, J. M. Rosenberg, O. B. Kallai, R. E. Dickerson, K. Itakura, A. D. Riggs, and K. S. Matthews, *J. Biol. Chem.* **255,** 10707 (1980).

[10] T. J. Daly and K. S. Matthews, *Biochemistry* **25,** 5479 (1986).

[11] R. B. O'Gorman, M. Dunaway, and K. S. Matthews, *J. Biol. Chem.* **255,** 10100 (1980).

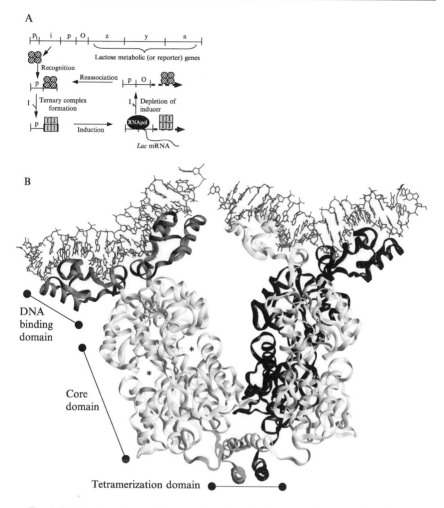

FIG. 1. Functional cycle and structure of LacI. (A) LacI function. Tetrameric LacI (striped circles) binds to its target operator sequence (O) and inhibits transcription of *lac* structural enzymes (z, y, and a). In the presence of inducer (I), LacI undergoes a conformational change (shaded squares) that is communicated to the DNA-binding site on the protein and results in release of operator. Thus, RNA polymerase transcription of downstream genes can occur. Once inducer is depleted, LacI resumes the conformation with high affinity for the operator site and resumes repression. (B) LacI structure. LacI is frequently described as a dimer of dimers. This structure shows each dimer bound to DNA (represented in stick form at top) [pdb code 1lbg; M. Lewis, G. Chang, N. C. Horton, M. A. Kercher, H. C. Pace, M. A. Schumacher, R. G. Brennan, and P. Lu, *Science* **271**, 1247 (1996)]. The right dimer is black and grey to highlight participating monomers. The left dimer is shaded to indicate the DNA-binding, core, and tetramerization domains. The asterisks indicate the positions of the two inducer-binding sites within a dimer.

SCHEME I

THERMODYNAMIC DESCRIPTION OF END POINTS AND BINDING EQUILIBRIA

Note: To determine appropriate concentrations of ligand and protein for each measurement, proper analysis of a new protein will require iterations of steps 1 and 2.

1. Determine stoichoimetry for each ligand.
   a. Concentration of either protein or ligand may be varied.
      i. The other species has fixed, precisely known concentration at 10- to 100-fold above K_d for binding. *Note:* These conditions ensure that, at low concentrations, all varied ligand is bound. At higher concentrations, the varied species should surpass the concentration of the constant species.
   b. Method must detect all specific binding events for the given ligand.
   c. Analyze binding stoichiometry.
      i. *Option 1:* Visually fit the pre- and postsaturation data with lines (Fig. 2A, inset). Determine their point of intersection. The x axis value of this intersect, divided by the known concentration of the constant species, yields the stoichiometric ratio.
      ii. *Option 2:* Use linear regression to determine the slope of the line for the presaturation data. This value is equal to the stoichiometric ratio.

Note: In our hands, both analyses provide similar results. Once stoichiometry is established, this process may be used to establish the active fraction for a given protein preparation.

2. Determine K_d for each ligand.
   a. Concentration of either protein or ligand may be varied.
      i. Concentration of remaining species is fixed >10-fold below K_d for the binding event monitored, but need not be precisely determined. *Note:* Derivation of Eqs. (1)–(3) relies on a number of assumptions that are possible only if the concentration of fixed species is at least 10-fold below K_d. Proper fitting with Eqs. (1)–(3) requires that the concentration range of the varied species is wide enough to establish well-defined pre- and posttransition baselines.
   b. Binding event(s) detected by method must be well defined. *Note:* The investigator must establish whether the technique simultaneously monitors all binding events of a given ligand or reports a specific event at a single site.
   c. Binding phenomena may be complex. *Note:* Several volumes in this series are dedicated to choosing appropriate mathematical models to describe biological processes, as well as to discussing issues related to parameter and error estimation. These include earlier volumes on the current topic (*Energetics of Biological Macromolecules,* Parts A–C, Volumes 259, 295, and 323) as well as the series titled *Numerical Computational Methods* (Volumes 210, 240, and 321).
   d. Equations for analysis of LacI equilibrium binding:

$$Y_{obs} = Y_{max} \frac{[V]}{K_d + [V]} + c \tag{1}$$

Equation (1) is appropriate to fit data for a binding event that does not have cooperativity and relies on a gain of signal as binding progresses (Fig. 2A). Here, Y_{obs} is the signal corresponding to the amount of bound complex, and Y_{max} is the observable property when 100% of the constant species is bound to the varied species, V. K_d is the equilibrium dissociation constant, and c is the background signal at low concentrations of V, when no ligand is bound.

(continued)

SCHEME I (continued)

$$Y_{obs} = Y_{max} - \left(Y_{max} \frac{[V]^n}{K_d^n + [V]^n} \right) + c \qquad (2)$$

Equation (2) is a variation of Eq. (1) for processes with cooperative binding (n is the Hill coefficient) and that detect a loss of signal as binding occurs (Fig. 2B). To normalize the data as fraction bound (R) for comparison of multiple experiments (Fig. 2), Y_{obs} is algebraically corrected with Y_{max} and c, so that

$$R = \frac{[V]^n}{K_d^n + [V]^n} \qquad (3)$$

Remember that $n = 1$ for noncooperative binding.

3. Binding methods. *Note:* The method chosen will be directed by the magnitude of K_d (tight binding events require more sensitive techniques; weak binding events require rapid separation techniques) and the availability of spectroscopic probes or radiolabeled ligands.
   a. Methods for DNA binding:
      i. Nitrocellulose filter binding of radiolabeled DNA[a]
      ii. Gel retardation of radiolabeled DNA[b]
      iii. Gel retardation using fluorescent dyes[c]
      iv. Fluorescence polarization of labeled DNA[d]
   b. Methods for IPTG binding:
      i. Equilibrium dialysis with radiolabeled IPTG (no longer commercially available)[e]
      ii. Ammonium sulfate precipitation assay with radiolabeled IPTG[e]
      iii. Fluorescence spectral shift.[f] *Note:* The validity of this method was confirmed by comparison with other assays and cannot be used to determine tetramer stoichiometry. It could be used to determine the fraction of monomers with active inducer-binding sites.

[a] I. Wong and T. M. Lohman, *Proc. Natl. Acad. Sci. USA* **90**, 5428 (1993).
[b] J. Carey, *Methods Enzymol.* **208**, 103 (1991).
[c] Molecular Probes, "BioProbes 42." Molecular Probes, Eugene, OR, 2003, p. 26.
[d] J. J. Hill and C. A. Royer, *Methods Enzymol.* **278**, 390 (1997).
[e] S. Bourgeois, *Methods Enzymol.* **21**, 491 (1971).
[f] S. L. Laiken, C. A. Gross, and P. von Hippel, *J. Mol. Biol.* **66**, 143 (1972).

Binding equilibria for LacI can be interpreted by two widely used models—the Monod, Wyman, and Changeux (MWC) model or the Koshland model.[9] Interestingly, LacI-binding curves can be effectively fit by either of these models. Using the MWC model to fit inducer-binding curves in the absence and presence of DNA, the allosteric constant (L) is ~1, indicating equal populations of the R and T states in the free protein.[9,10] Efforts to isolate partially liganded intermediates in the LacI allosteric change in response to inducer are frustrated by the rapidity of this shift. These states appear to occur only transiently as the protein changes conformation. Alternatively, molecular dynamics simulations allow interpolations to assess these intermediates (see below).

SCHEME II

ASSESSING ALLOSTERY

1. Does one ligand affect the free energy of binding for the other ligand(s)?
   a. Does ligand A bind with different affinities in the presence/absence of ligand B?
2. Is the process reciprocal? *Note:* Thermodynamic cycle should be additive.
   a. If not, a binding event is undetected, or the method is affected by processes other than binding.
3. Experimental design. *Note:* Reaction contains protein and two ligands.
   a. Measure K_d for each ligand in the presence of a fixed concentration of other ligand.
      i. The fixed ligand should be under stoichiometric conditions (Scheme I, step 1). Note that the concentration of this ligand should be 10- to 100-fold greater than the highest K_d for this ligand. *Note:* This is to assure that the protein remains bound to the fixed ligand under all current experimental conditions.
      ii. Concentrations of the remaining ligand and protein should be under equilibrium conditions (Scheme I, step 2), that is, either protein or ligand should have a concentration 10- to 100-fold below the lowest K_d for this ligand.
      iii. Successful execution of these experiments again requires iteration to determine the correct concentrations for each species.

   Example 1: DNA affinity in the presence of inducer. LacI binds inducer IPTG with a K_d of $\sim 10^{-6}$ M in the absence of operator DNA, but this value increases to $\sim 10^{-5}$ M in the presence of operator DNA. To determine K_d for DNA in the presence of inducer, IPTG concentration must be at least 10^{-4} M and preferably higher. LacI concentration is varied, and operator DNA must be at most 10^{-12} M (10-fold below the K_d for DNA binding in the absence of inducer).

   Example 2: Inducer affinity in the presence of DNA. LacI binds operator with a K_d of $\sim 10^{-11}$ M in the absence of inducer, but this value increases to $\sim 10^{-8}$ M in the presence of inducer. To determine binding affinity for inducer in the presence of operator DNA, the concentration of operator must be at least 10^{-7} M. In this case, inducer concentration is varied, and LacI concentration must be at most 10^{-7} M (10-fold below K_d for IPTG in the absence of operator).

4. Complex analysis using a global model, such as the Monod–Wyman–Changeux model. *Note:* Binding measurements should be repeated for each ligand while varying the concentration of fixed ligand.
   a. The concentrations of fixed ligand will not be at saturating levels. They should range between 10-fold higher and 10-fold lower than the K_d for binding the fixed ligand.
   b. Special care should be taken to ensure that solution conditions are the same in all experiments, including the different binding assays for the two ligands.[a,b]

---

[a] R. B. O'Gorman, J. M. Rosenberg, O. B. Kallai, R. E. Dickerson, K. Itakura, A. D. Riggs, and K. S. Matthews, *J. Biol. Chem.* **255,** 10107 (1980).

[b] T. J. Daly and K. S. Matthews, *Biochemistry* **25,** 5479 (1986).

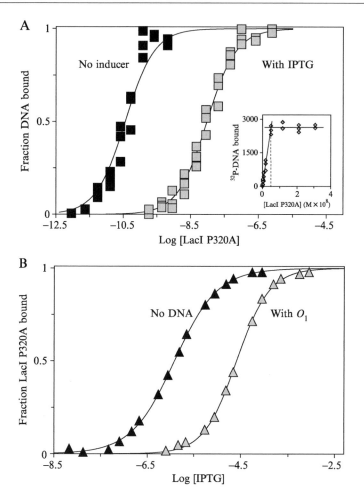

FIG. 2. Allostery in LacI. Inducer and operator binding curves in (A) and (B) demonstrate the reciprocal influence of inducer and operator on binding affinity for the LacI mutant P320A. Allosteric response is demonstrated by lower operator affinity in the presence of inducer, and vice versa. Note that inducer binding in the presence of operator DNA (gray triangles) is cooperative (steeper slope of transition). The *inset* in (A) demonstrates stoichiometry of operator binding. For a single experiment in which DNA binding is monitored by the nitrocellulose filter binding method (A), data points for each concentration are determined in triplicate. In contrast, data from a single fluorescence experiment to measure IPTG binding (B) are expressed as a single point for each concentration.

Intermediate Processes in Mechanisms of DNA Binding

Thermodynamic analysis can provide significant insight into mechanisms of protein action. For LacI, Spolar and Record[12] observed large values of ΔC_p that they ascribed to protein folding coupled to DNA binding (see Record et al.[4] for details of measurements). This result was consistent with early nuclear magnetic resonance (NMR) experiments[13] demonstrating that the N-terminal domain, known by genetic studies to be essential for DNA binding, was highly mobile in the absence of DNA. Confirmation was provided by both NMR and crystallographic structural studies demonstrating that the hinge region between amino acids 50 and 60 folds and inserts into the minor groove of DNA and appears to be unfolded in the free protein (Fig. 4A).[14,15] However, the structural information also demonstrated that ΔC_p measured for LacI appears larger than can be accounted for by burial of surface residues on hinge helix folding. Further studies will be required to sort out these detailed issues.

Subunit Contributions to Binding and Allostery

Because binding and assembly can be energetically linked processes, oligomeric proteins pose unique problems in thermodynamic analysis. The use of specifically modified proteins can be useful in parsing subunit contributions to function. For example, two variants of LacI were successfully used to measure intrinsic binding affinities, to establish linked equilibria between assembly and binding (Fig. 3), and to limit the range of the pathway for allosteric communication. First, intrinsic inducer binding was illuminated using the monomeric LacI mutant, Y282D (Fig. 4A).[16] This mutant retains the inducer-binding properties of the wild-type protein, but does not bind to DNA. Second, a C-terminal deletion of LacI results in the formation of dimer (Fig. 4A).[17]

As shown in Fig. 1B, LacI is assembled as a dimer of dimers, and the C-terminal leucine heptad repeat sequence (the final 18 amino acids) forms the primary dimer–dimer interface. Note that this assembly domain was identified by mutation and deletion of segments of the C terminus well

[12] R. S. Spolar and M. T. Record, Jr., Science 263, 777 (1994).
[13] N. Wade-Jardetzky, R. P. Bray, W. W. Conover, O. Jardetzky, N. Geisler, and K. Weber, J. Mol. Biol. 128, 259 (1979).
[14] M. Lewis, G. Chang, N. C. Horton, M. A. Kercher, H. C. Pace, M. A. Schumacher, R. G. Brennan, and P. Lu, Science 271, 1247 (1996).
[15] C. A. E. M. Spronk, G. E. Folkers, A.-M. G. W. Noordman, R. Wechselberger, N. van den Brink, R. Boelens, and R. Kaptein, EMBO J. 18, 6472 (1999).
[16] T. J. Daly and K. S. Matthews, Biochemistry 25, 5474 (1986).
[17] J. Chen and K. S. Matthews, J. Biol. Chem. 267, 13843 (1992).

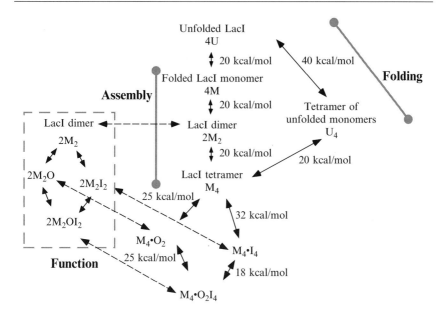

FIG. 3. Linked thermodynamic equilibria in LacI function, assembly, and unfolding. This schematic is a simplified thermodynamic description of LacI behaviors. M, LacI monomer; O, operator DNA; I, inducer IPTG. DNA and inducer binding equilibria between the fully liganded end points are represented for dimer *(left)* and tetramer *(bottom, center)*. These processes are linked to assembly *(top, center)* and folding *(right)*. The values by each arrow correspond to the Gibbs free energy of each transition.

before the structure was solved.[17–19] Further, the inducer-binding parameters, including cooperativity in the presence of operator DNA, are identical for dimer derivatives and for tetramer. Thus, the pathway for communication of cooperative inducer binding (and all intermediates along this path) must reside within the dimer. Similarly, the dimer exhibits the same allosteric response to inducer binding as tetramer, so that the physical pathway for allosteric communication to the DNA-binding site must also exist within the dimer rather than between dimers. However, deletion of the C terminus revealed that the monomer–monomer affinity is lower than the dimer–DNA affinity; hence, the dimeric mutations exhibit linked thermodynamic equilibria for assembly and binding (Fig. 3).[20] After

[18] A. E. Chakerian, V. M. Tesmer, S. P. Manly, J. K. Brackett, M. J. Lynch, J. T. Hoh, and K. S. Matthews, *J. Biol. Chem.* **266,** 1371 (1991).
[19] S. Alberti, S. Oehler, B. von Wilcken-Bergmann, H. Krämer, and B. Müller-Hill, *New Biol.* **3,** 57 (1991).
[20] J. Chen and K. S. Matthews, *Biochemistry* **33,** 8728 (1994).

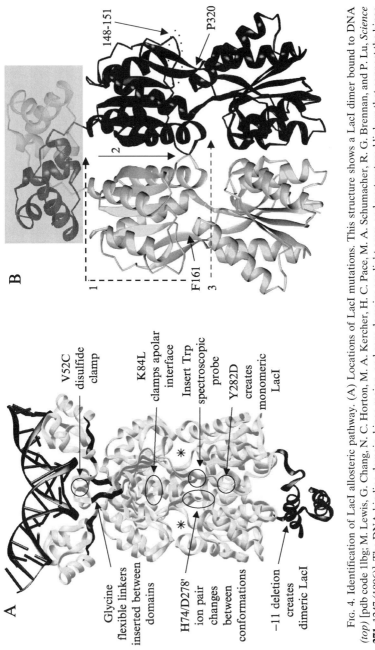

Fig. 4. Identification of LacI allosteric pathway. (A) Locations of LacI mutations. This structure shows a LacI dimer bound to DNA (*top*) [pdb code 1lbg; M. Lewis, G. Chang, N. C. Horton, M. A. Kercher, H. C. Pace, M. A. Schumacher, R. G. Brennan, and P. Lu, *Science* **271,** 1247 (1996)]. The DNA-binding domain, hinge region, and core domains are light gray, and unstructured linkers that connect the hinge to the DNA-binding and core domains are black. The tetramerization domain (*bottom*) is also colored black. Circles indicate the positions of various mutations discussed in text. (B) Allosteric pathway illuminated by TMD. This structure models a LacI dimer with a truncated C terminus in the DNA-bound conformation [pdb code 1efa; C. E. Bell and M. Lewis, *Nat. Struct. Biol.* **7,** 209 (2000)]. One monomer is

accounting for this linkage, detailed DNA-binding measurements demonstrate that the intrinsic affinity of a dimer for operator is similar to tetramer.[20] These experiments also demonstrated the necessity of examining contributions from assembly to LacI function.

Intermediates in LacI Assembly and Folding

Folding and assembly of oligomeric proteins can also be examined by thermodynamic approaches, even in the absence of detailed structural information. The LacI tetramer is highly stable, and no monomeric or dimeric intermediates are detected in purified wild-type protein. Therefore, the dimer deletion mutants and Y282D monomer mutant were critical to determine the energetics for assembly of this protein. When assembly is linked to unfolding, the concentration of denaturant required for unfolding is dependent on the concentration of the protein. Denaturing tetrameric LacI to four unfolded monomers should evince a high degree of concentration dependence. Surprisingly, this dependence was not observed for wild-type tetrameric LacI.[21] However, data from denaturing the dimeric mutant proteins did demonstrate a protein concentration dependence, indicating that two monomers were in equilibrium with dimer.[20] Together, these studies suggested that disruption of the monomer–monomer interface and monomer unfolding occurred before disruption of the C-terminal leucine heptad repeat structure that holds the two dimers together.[21] Data were consistent with the dissociation/unfolding of the monomeric units within the tetramer to produce an intermediate structure in which these unfolded monomeric units were tethered at the C terminus. Using denaturation data for monomer, dimer, and tetramer variants of LacI, the individual thermodynamic contributions of monomer folding, dimer assembly, and tetramer

---

[21] J. K. Barry and K. S. Matthews, *Biochemistry* **38,** 6520 (1999).

---

colored light gray and the other is dark gray. Arrows indicate the pathways of changing interactions identified by TMD during the LacI conformational change from the DNA- to inducer-bound structure. Note that the simulation did not include the DNA-binding domains (shadowed at *top*). Changes associated with pathway 1 (long-dash line) were initiated by differences at amino acid 149 (dotted circle), which contacts inducer, and conclude at K84. Pathway 1 occurred on only one monomer. Changes in the monomer–monomer interface of pathway 2 (solid line) involved both subunits; K84 is an integral participant. Subsequently, the core domain of the second monomer underwent structural changes similar to pathway 1. Changes along pathway 3 (short-dash line) occurred last in the simulation and involved both monomers. Pathway 3 originates in the core pivot (centered around F161) and included π stacking between the H74 rings of both monomers (T. C. Flynn, L. Swint-Kruse, Y. Kong, C. Booth, K. S. Matthews, and J. Ma, *Protein Sci.* **12,** 2523 (2003)).

assembly could be determined (Fig. 3). Studies with highly stable variants of LacI (see below) were used to confirm this arrangement.[21]

Contributions to Function from Small Protein Regions

As evident from the foregoing discussion, thermodynamic measurements can provide significant insight into protein functional components, even when a structure has not yet been determined for a given protein. However, when available, structural knowledge provides a rationale for the targeted design of mutants to test hypothetical mechanisms. Thermodynamic binding analyses of such mutants (including stoichiometry, equilibrium dissociation constants, and allosteric response) provide a mechanism to limit and/or define the role of specific regions in the function of a protein. To design relevant mutants, detailed examination of the structure(s) of the protein and comparison to structures of similar proteins can be helpful. The goal of the mutation can be to (1) disrupt existing interactions that are expected to be critical, (2) place spectroscopic reporters in different regions, (3) block communication by introducing flexibility, or (4) insert a molecular "clamp" to interrupt the allosteric pathway (Scheme III). An additional, powerful tool utilizes random mutagenesis and phenotypic screening to address similar issues (Scheme IV). An example of each of these strategies follows.

Testing Interactions Critical to Allosteric Response

Detailed, thermodynamic analysis of point mutations can be used to distinguish the role of specific side chains in function and to test hypotheses based on static structures of end-point conformations. In the structures of LacI, H74 and D278' (a cross-monomer interaction; Fig. 4A) form an electrostatic interaction only in the inducer-bound form of the protein. On the basis of the differences for these residues between alternate conformations, this ion pair was postulated to be important in the allosteric pathway.[14] To test this hypothesis, individual substitutions were made, and the thermodynamic properties of the altered proteins were measured.[22] Although apolar amino acids at position 74 altered ligand-binding properties, introduction of a negatively charged residue (H74D) had minimal effect, demonstrating that the salt bridge could not be essential for stabilizing or destabilizing the inducer-bound conformation. Overall, the individual roles played by H74 and D278 appeared more important in LacI function than a direct contact between them.[22] Thus, the hypothesis formed from structural analysis alone was incorrect.

[22] J. K. Barry and K. S. Matthews, *Biochemistry* **38**, 3579 (1999).

SCHEME III

DESIGNING MUTANTS

1. No structural information:
   a. Examine sequence for homology to other proteins.
      i. Delete domains (e.g., LacI–11).
      ii. Change residues conserved in function or assembly.
   b. Utilize prediction algorithms for predicting secondary structure.
      i. Change residues so that structural elements are interrupted or enhanced.
      Example: A proline substitution should abolish an α helix, whereas an alanine substitution might favor the helix for a region with a putative helix–coil transition.
2. Structural data available:
   a. Interrupt possibly critical interactions:
      Direct contact to ligand
      Direct contact between domains or subunits (e.g., LacI Y282D)
      Significant change between structures of two conformations (e.g., LacI H74D)
   b. Insert spectroscopic probe.
      i. Choose region that might be sensitive to change.
      ii. Make substitutions. Possible substitutions are as follows:
         Tryptophan: Change in fluorescence signal, polarization, lifetime, or quenching on ligand binding (e.g., LacI F293W)
         Cysteine: Attach spectroscopic probe via disulfide bond (reversible) or other agents targeted to sulfhydryl moiety
   c. Add flexibility between domains.
      i. Make glycine linkers (e.g., LacI Q60G + 3).
      ii. Substitute prolines with glycine/alanine (e.g., LacI P320A).
   d. Add clamp between domains or subunits.
      i. Add disulfide bond (e.g., LacI V52C).
      ii. Increase hydrophobicity to decrease mobility of interface (e.g., LacI K84L).
      iii. Add bulky side chain to sterically hinder alternate conformations (e.g., LacI H74W).
      iv. Make proline substitution to stiffen extended chains/loops (e.g., LacI S151P).

Tryptophan Scanning: Introducing a Spectroscopic Probe

Tryptophan accounts for the majority of fluorescence in most proteins containing this amino acid and is frequently present only in a few sites. Because indole fluorescence reports on its nearby environment, substitution of this residue at selected sites can illuminate local changes. Elimination of wild-type tryptophan sites and substitution at selected sites provides unique spectroscopic information within the structure. Binding analyses coupled with fluorescence alterations in the presence of ligands, polarization data, and fluorescence lifetime data can yield information useful in interpreting conformational shifts and local environment. For example, in LacI the fluorescence and quenching properties of single tryptophan residues placed in the central region of the protein (Fig. 4A) were affected by binding of one or both ligands.[23] This result indicates

SCHEME IV

LET THE SYSTEM DO THE WORK: RANDOM MUTAGENESIS

1. Select the target function.
   Examples: Assembly, ligand specificity (protein–protein, protein–small molecule, protein–DNA), allosteric response to alleviate transcription repression
2. Design an *in vivo* screen.
   a. Promoter-reporter in bacteria, yeast, or eukaryotic cells
      i. Protein–ligand (protein, small molecule, or DNA) effects on transcription
   b. Yeast two-hybrid assay
      i. Protein–protein interactions
   c. Phage display assay
      i. Protein–protein, protein–ligand interactions
   d. Variables to manipulate
      i. DNA copy number/promoter strength for protein of interest
      ii. Promoter strength for reporter gene
      iii. Number of DNA-binding sites to control reporter gene
      iv. Concentration of small molecule
      v. Sensitivity of assay, using different substrates: colorimetric, fluorescent, radioactive
      vi. Differential growth and detection conditions
      vii. Varied solution conditions for phage display
   e. Factors that affect *in vivo* screening/biophysical comparison
      i. Protein expression levels
      ii. Protein degradation
      iii. Assay repeatability can be poor; several rounds of screening are required for each positive.
      iv. *In vivo* conditions may not match *in vitro* solution conditions

Example: LacI assembly. LacI binds DNA only when assembled to at least a dimer. Hence a promoter–reporter assay can utilize DNA binding and subsequent transcription repression of downstream genes to detect assembly. The mutation Y282D produces a monomeric construct that cannot bind DNA. Random mutants of Y282D were screened. At least 22 second-site mutations restore *in vivo* repression, presumably through compensating the assembly defect of Y282D.[a]

Cell strain	Promoter	Reporter	Substrate	Protein	Result
DH5α bacteria	Lac O_1	β-Galactosidase on high-copy plasmid pZCam	X-Gal	Wild-type LacI (positive control for repression)	White colony
				Monomeric Y282D (positive control for colorimetric assay)	Blue colony
				Randomly mutated Y282D LacI on high-copy pAC1	Search for white colonies among blue colonies

Example: LacI inducer specificity. LacI DNA binding is diminished in the presence of inducer. Thus, LacI mutants that can be induced by other sugars or low levels of IPTG can be identified by loss of repression under these conditions. Random mutants of LacI were screened against 10 possible inducers in a promoter–reporter assay. Two mutants were identified as being induced by all sugars tested, including 10-fold less IPTG.[b]

Cell strain	Promotor	Reporter	Substrate	Protein	Result	
					No inducer	Varied inducer sugars
Bacterial strain 3.300	Lac O^1	Single copy, genomic β-galactosidase	MUG	Wild-type LacI (positive control for repression; positive control for IPTG induction; negative control for varied sugar induction)	White colony	IPTG: fluorescent colony Other sugars: white colony
				Monomeric Y282D (positive control for fluorescence assay)	Fluorescent colony	Fluorescent colony
				Randomly mutated LacI on low-copy pCRE	Repression intact: white colony Repression lost: fluorescent colony	Search for fluorescent colonies with other sugars that can repress without inducer

[a] L. Swint-Kruse, C. R. Elam, J. W. Lin, D. R. Wycuff, and K. S. Matthews, *Protein Sci.* **10**, 262 (2001).
[b] L. Swint-Kruse, H. Zhan, B. M. Fairbanks, A. Maheshwari, and K. S. Matthews, *Biochemistry* **42**, 14004 (2003).

that structural regions far from the N-terminal DNA-binding domain participate in forming the conformation with high affinity for DNA.

Disrupting Communication: Additional Flexibility

Glycine-rich regions can be used to introduce flexibility in segments of the structure or to increase the relative mobility between protein domains. The goal in LacI was to uncouple regions that transmit the allosteric message—specifically the N-terminal DNA-binding domain and the core inducer-binding domain (Fig. 4A)—by substituting and inserting additional glycines in the hinge helix that connects these domains.[24] In LacI, Q60G slightly enhanced DNA-binding affinity but had no impact on inducer binding or allosteric response. When additional glycine residues were inserted C terminal to Q60G, the affinity of LacI for O_1 operator DNA was affected. Inducer-binding affinity was little changed, indicating that the core domain was structurally intact. However, the allosteric impact of inducer on O_1 binding diminished as a function of the number of glycines. Therefore, the allosteric pathway must have been interrupted; presumably, added flexibility disrupted the interface between the two functional domains. Interestingly, this set of LacI variants provided the first illustration that another ingredient is essential for communication between the two domains, because the allosteric response could be restored by changing the operator DNA sequence.[24] Thus, ligand variation can be useful in understanding a function previously thought to be exclusive to the protein. Specifically, DNA cannot be assumed to be a "neutral" target.

Disrupting Communication: Introducing a Clamp

"Clamping" is another approach to uncoupling regions involved in transmitting allosteric response. For LacI, the side chains of Val-52 and -52' were identified as the site of closest apposition between the two hinge helices that insert into the minor groove of operator DNA (Fig. 4A). The mutant V52C was generated to introduce the potential for a disulfide bond under oxidizing conditions, that is, a clamp to maintain juxtaposition of these two helices.[25] Control experiments with V52C under reducing conditions indicated that mutant properties were similar to wild type. However, under oxidizing conditions, V52C exhibited enhanced affinity for O_1 operator DNA and wild-type affinity for inducer. Of greatest interest, when the disulfide clamp was formed, O_1 operator binding was completely

[23] J. K. Barry and K. S. Matthews, *Biochemistry* **36,** 15632 (1997).
[24] C. M. Falcon and K. S. Matthews, *Biochemistry* **39,** 11074 (2000).
[25] C. M. Falcon, L. Swint-Kruse, and K. S. Matthews, *J. Biol. Chem.* **272,** 26818 (1997).

insensitive to inducer binding, indicating disruption of the LacI allosteric pathway.[25]

An interesting component of the thermodynamic analysis was the observation that ΔC_p for the oxidized V52C LacI was essentially zero, suggesting that hinge folding was no longer coupled to DNA binding.[26] Apparently, this mutant may approximate intermediates in the allosteric pathway—the structural changes may proceed partway before reaching a blockade. Interestingly, allosteric communication in oxidized V52C could again be restored with alternate operator sequences,[26] further emphasizing the role of DNA sequence in LacI allostery. These results illustrate that parsing and characterizing the energetic components of protein function can be unexpectedly complex, even with a series of crystallographic and NMR structures and established binding analyses. Understanding the mechanisms by which small changes in ligand or protein alter function provides significant insight into the mechanistic details of allosteric response.

The hinge region is intimately involved in the DNA-binding process and indeed undergoes folding to generate the DNA-bound form of the protein. However, residues distant from both binding sites can exert strong influence on allosteric properties. In LacI, the monomer–monomer interface of the core domains is different for the DNA- and inducer-bound conformations.[14] The structural data helped explain the previously observed phenomenon that apolar substitutions for K84 (Fig. 4A) resulted in substantial stabilization of the tetrameric structure and slower kinetics of inducer binding.[27] Indeed, K84L is able to bind isopropyl-β-D-thiogalacto-pyranoside (IPTG) in >6 M urea,[21] and this mutant was used to parse subunit contributions from the overall unfolding of LacI (see above). Interestingly, the structure of K84L has been determined and shows substantial rearrangement in the subunit interface.[28]

Together, the results indicate that in wild-type protein, the positive lysine side chain serves as a "hair trigger" in the monomer–monomer interface: K84 destabilizes the otherwise hydrophobic interface of the DNA-bound form so that LacI is poised to respond to inducer binding. During the allosteric response of wild-type protein, the K84 side chain moves out of the interface, and more apolar interactions are formed. The K84L substitution increases interface hydrophobicity, effectively removing one key driving force of the allosteric conformational change (see below for details from targeted molecular dynamics). K84L thus clamps the subunit interface in the DNA-bound conformation.

[26] C. M. Falcon and K. S. Matthews, *Biochemistry* **40**, 15650 (2001).
[27] W.-I. Chang, J. S. Olson, and K. S. Matthews, *J. Biol. Chem.* **268**, 17613 (1993).
[28] C. E. Bell, J. Barry, K. S. Matthews, and M. Lewis, *J. Mol. Biol.* **313**, 99 (2001).

Random Mutagenesis: Letting the System Do the Work

The strategies presented above are best implemented when structural data are available to allow rational design of mutants. However, new functional phenotypes may be discovered independent of structural data (Scheme IV). This method allows the system itself to illuminate the regions critical to function, circumventing the problem of either a missing structure or the investigator's limited understanding of a given conformation. To illustrate, a promoter/reporter construct can be used to monitor the repression function of LacI in bacterial colonies. The monomeric Y282D variant cannot bind DNA, but phenotypic screening for *in vivo* repression of randomly mutated Y282D identified mutants that restore DNA binding. Some of these revertants were in anticipated regions of the protein (i.e., subunit interfaces), whereas others appeared at sites that were not intuitive from studies of static structures.[29]

Random studies may be adjusted to target different functions. For example, an independent screen designed to identify LacI mutants induced by alternate sugars yielded two variants, L148F and P320A.[30] Both sites are located in the solvent-exposed region surrounding position 150 in LacI (Fig. 4B), which is well separated from the DNA-binding site and just outside the inducer-binding site. Interestingly, this region contains several second-side revertants for Y282D (e.g., S151P).[29] Therefore, results of the two *in vivo* screens, coupled with thermodynamic analyses, converge to suggest that this "core pivot" region plays an essential role in the allosteric pathway of LacI.[30]

Atomic-Level Intermediates: Molecular Dynamics Simulations

A common frustration is that many intermediate states are not populated at thermodynamically detectable levels or that partially liganded intermediates cannot be trapped. However, information about intermediates and pathways may be obtained by molecular dynamics (MD) simulations to illuminate potential motions of a protein. Scheme V provides an outline for undertaking a molecular dynamics analysis. For a protein the size of LacI, molecular dynamics of the intact system is not feasible; however, various domains may be examined. For LacI, motions that occur in the N-terminal DNA-binding domain on dissociation of DNA were simulated with MD. Results indicated that the helix–turn–helix

[29] L. Swint-Kruse, C. R. Elam, J. W. Lin, D. R. Wycuff, and K. S. Matthews, *Protein Sci.* **10,** 262 (2001).
[30] L. Swint-Kruse, H. Zhan, B. M. Fairbanks, A. Maheswari, and K. S. Matthews, *Biochemistry* **42,** 14004 (2003).

SCHEME V

MOLECULAR DYNAMICS

1. High-resolution structural coordinates are required.
   a. Molecular dynamics (MD) simulation requires one structure.
   b. Targeted molecular dynamics (TMD) simulation requires two structures.

2. Edit pdb to requirements of simulation program.
   a. Add hydrogen atoms to X-ray structures (not necessary for NMR structures).
   a. Delete regions not relevant and/or that make system too large for simulation.
       i. System size depends on simulation algorithm, computing power, total number of atoms, and length of time researcher is willing to commit for obtaining results.
       ii. *Remember:* Deletions may have experimentally relevant implications in simulation result.
   c. Atoms in start and finish structures for TMD must be identical.
       i. Ligand must be deleted in simulation of ligand-bound to free structure.

3. Solvate structure.
   a. Explicit solvent waters are best.
       i. *Option 1:* Solvent box: Vary shape and protein position.
       ii. *Option 2:* Solvent shell.
   b. Explicit counterions versus ionic atmosphere.
       i. Explicit ions will slow simulation.
       ii. A small number of experimentally important ions may provide compromise.

4. Perform equilibration and energy minimizations to accommodate solvent.
   a. *Remember:* If anything was deleted from original structure, effects may begin to show up here. To check, solvate and equilibrate with/without deletions.

5. Run trajectory.

6. Analyze data. *Note:* Tools and strategies outlined below are also useful for comparing structures of homologous proteins.
   a. Visually watch "movie" of trajectory.
   b. Select structures at representative time points.
       i. Align with starting structure, use multiple domains/regions as basis. Web tool: Combinatorial Extension will align structures using regions $\geq$16 amino acids as the reference (http://cl.sdsc.edu/ce.html).[a] Edit pdb Chain ID and SEQRES to maintain quaternary structure.[b]
       ii. Inspect trajectory structures for changes, define target regions.
   c. Compare atom contacts in target regions.
       i. Web tool: Contacts of Structural Units will generate detailed lists of contacts (http://bioinfo.weizmann.ac.il:8500/oca-bin/lpccsu).[c]
   d. Plot variable versus trajectory time:
       i. ψ, φ, other side-chain angles
       ii. Atom–atom distance
       iii. Solvent accessibility
   e. Follow changes through logical connections. This may lead to unnoticed areas. *Example:* If changes are observed around amino acid X, assess whether the distance varies between X and Y side chains. If so, monitor changes in contacts to amino acid Y. If distance or angle changes occur between atoms of residues Y and Z, examine Z behavior. Repeat until no new changes are detected.
   f. *Remember:* Explicit water molecules or solvent ions may be direct participants in protein dynamics and may be integral to analysis.

[a] I. N. Shindyalov and P. E. Bourne, *Protein Eng.* **11**, 739 (1998).
[b] L. Swint-Kruse, C. Larson, B. M. Pettitt, and K. S. Matthews, *Protein Sci.* **11**, 778 (2002).
[c] V. Sobolev, A. Sorokine, J. Prilusky, E. E. Abola, and M. Edelman, *Bioinformatics* **15**, 327 (1999).

and hinge regions within LacI (Fig. 4A) are motionally independent.[31,32] This result suggests that multiple structures are possible in the absence of DNA; that is, intermediates following inducer binding are likely to be ensembles instead of single conformations. This information is useful in understanding the potential steps along the allosteric pathway and indicates that the products will be diverse rather than uniform.

For larger proteins, a specific form of molecular dynamics may also be useful—targeted MD (TMD).[33] In this case, two end-point structures are required, and the TMD forces the structure to change from one conformation to the other. This analysis provides the opportunity to follow linked changes through the structure, in a sense providing an atom-by-atom series of kinetic intermediates. TMD results for the LacI core domain provide a detailed allosteric pathway that explains the surprising results from genetic screens surveying random mutants (see above; Fig. 4B).[34] Specifically, motions in the core pivot region are central to the conformational change that occurs in the DNA-bound structure in response to inducer binding. Therefore, substitutions in the core pivot potentially favor one conformation of the protein over the other and thereby alter functional properties. Interestingly, the TMD pathway also indicates asymmetric changes in the symmetric homodimer that are requisite for experimentally observed cooperativity of inducer binding of the LacI–DNA complex.[9,10] This "theoretical" treatment provides information about the allosteric pathway that can be used to design new mechanisms to interrupt the pathway and thereby understand the ways in which information is transmitted through the LacI structure.

Conclusion

Thermodynamic analysis of allosteric and cooperative proteins is an essential first step to deciphering their functional mechanisms. This process provides crucial baseline data for comparing differences elicited by either altered solution conditions or protein modification. Data describing binding, assembly, and folding can then be combined with molecular genetics—whether domain deletion, targeted mutagenesis, or random mutagenesis—to test specific hypotheses and narrow the allosteric pathway. When structural information is available, designed variants and molecular dynamics expand the range of questions that can be addressed and the detail that can be discerned.

[31] L. Swint-Kruse, K. S. Matthews, P. E. Smith, and B. M. Pettitt, *Biophys. J.* **74,** 413 (1998).
[32] L. Swint-Kruse, C. Larson, B. M. Pettitt, and K. S. Matthews, *Protein Sci.* **11,** 778 (2002).
[33] J. Schlitter, M. Engels, P. Krüger, E. Jacoby, and A. Wollmer, *Mol. Simul.* **10,** 291 (1993).
[34] T. C. Flynn, L. Swint-Kruse, Y. Kong, C. Booth, K. S. Matthews, and J. Ma, *Protein Sci.* **12,** 2523 (2003).

Well-designed thermodynamic analyses lie at the very heart of under-standing any allosteric or cooperative protein. This informative avenue is feasible is even when detailed structural information is not available—a situation that is increasingly common as proteins with unstructured regions are identified and purified. Their structural flexibility impedes crystallization, while their size frequently exceeds the limitations of NMR analysis. In these proteins, thermodynamic and genetic analyses are therefore the most profitable pathways to functional insight.

Acknowledgments

We thank Dr. Sarah Bondos and Michelle Calabretta for feedback and comments on this manuscript. Hongli Zhan graciously provided the data for Fig. 2. The allosteric pathway simulated by TMD and presented in Fig. 4B was determined in collaboration with Terence C. Flynn and Dr. Jianpeng Ma (Rice University and Baylor College of Medicine). Support for the work reported from the Matthews Laboratory was from the NIH (GM22441) and the Robert A. Welch Foundation (C-576); the work of L.S.K. was supported in part by a training fellowship from the National Library of Medicine (LM07093).

[12] Linked Equilibria in Biotin Repressor Function: Thermodynamic, Structural, and Kinetic Analysis

By DOROTHY BECKETT

Introduction

Regulation of transcription initiation involves assembly of multisubunit protein complexes at the transcription control regions of genes.[1] To ensure that transcription initiation occurs in response to appropriate cellular conditions and with proper timing the assembly of these regulatory complexes is subject to control. Thus, many transcription regulators are allosteric proteins. As such their affinities for specific DNA sites are modulated either by small ligand binding or posttranslational modification. Those regulated by small ligand binding include the nuclear receptors in eukaryotes[2] and a large number of prokaryotic transcription regulators including catabolite activator protein (CAP),[3] and the purine[4] and diphtheria toxin[5]

---

[1] E. Martinez, *Plant Mol. Biol.* **50,** 925 (2002).
[2] A. C. Steinmetz, J. P. Renaud, and D. Moras, *Annu. Rev. Biophys. Biomol. Struct.* **30,** 329 (2001).
[3] S. Busby and R. H. Ebright, *J. Mol. Biol.* **293,** 199 (1999).
[4] F. Lu, R. G. Brennan, and H. Zalkin, *Biochemistry* **37,** 15680 (1998).

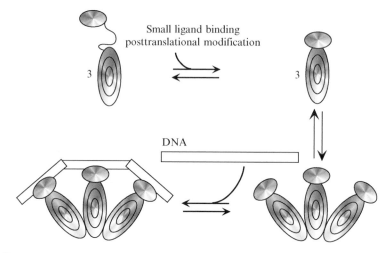

Fig. 1. Functional control of a transcription regulatory protein: Ligand binding or posttranslational modification induces conformational/dynamic changes that influence the self-assembly to trimer and/or DNA-binding properties of the protein.

repressors. The small molecule effectors for these proteins include steroids and nonsteroidal hormones, nucleotides and metal ions. Transcription regulators that respond to posttranslational modification include the SMAD proteins, which are involved in the response to transforming growth factor β (TGF-β) binding to cell surface receptors,[6] and the response regulators, which contribute to two-component signaling in prokaryotes and some lower eukaryotes.[7] In both of these examples the posttranslational modification that regulates function is phosphorylation. Small molecules and posttranslational modifications can be considered the primary effectors of the function of these proteins and the physiological function that is ultimately regulated is transcription initiation. However, the pathway to the final biological readout can be complex. It is common for the primary effectors to influence transcription initiation via their affects on both the protein assembly and DNA-binding properties of the transcriptional regulator (Fig. 1). Thus, in attempting to understand the functional regulation of these proteins, the following three questions must be addressed.

[5] M. M. Spiering, D. Ringe, J. R. Murphy, and M. A. Marletta, *Proc. Natl. Acad. Sci. USA* **100,** 3808 (2003).

[6] B. Y. Qin, S. S. Lam, J. J. Correia, and K. Lin, *Genes Dev.* **16,** 1950 (2002).

[7] A. H. West and A. M. Stock, *Trends Biochem. Sci.* **26,** 369 (2001).

1. How does binding of a small molecular effector by or posttranslational modification of the allosteric regulatory protein change its structural and/or dynamic properties?

2. How are the changes effected by small ligand binding or posttranslational modification propagated to alter the assembly properties of a transcriptional regulatory protein?

3. How is the occupancy of a regulatory site on DNA influenced by combined small ligand binding or posttranslational modification and protein assembly?

In addressing these questions methods must be developed to measure the energetics of small ligand binding, the self-assembly properties of the regulatory proteins, and its the DNA-binding properties. Moreover, because these processes are linked, elucidation of their functional regulation requires measurements of the individual interactions and their pairwise coupling. At present there are few systems for which this analysis has been accomplished. In this chapter the detailed studies of linked equilibria in the biotin repressor system are described. The approaches utilized in this system are directly transferable to other allosteric transcription regulatory proteins.

The multifunctional biotin repressor is an example of an allosteric DNA-binding protein. A summary of its functions is provided in Fig. 2. The protein is responsible for two functions in *Escherichia coli*.[8] First, it catalyzes the covalent linkage of biotin to the sole biotin-dependent carboxylase in *E. coli*, acetyl-CoA carboxylase and second, it represses transcription initiation at the biotin biosynthetic operon[9] (Fig. 3). The switch between the two functions is dictated by the cellular demand for biotin.[10] Posttranslational addition of biotin to a transcarboxylase occurs in two steps; the first in which ATP and biotin are utilized to form biotinyl-5'-AMP and the second in which the biotin moiety is transferred to a single lysine residue on the biotin carboxyl carrier protein (BCCP) subunit of the transcarboxylase.[11] In addition to its role in the biotin transfer reaction the adenylated form of biotin is the allosteric activator for site-specific binding of the repressor to the 40-base pair biotin operator.[12] In efforts to understand assembly of the transcription repression complex thermodynamic and kinetic methods to measure small ligand binding, protein assembly, and DNA binding in this system have been developed. Some of these methods and results of their application to this system are described

[8] J. E. Cronan, Jr., *Cell* **58**, 427 (1989).
[9] D. F. Barker and A. M. Campbell, *J. Mol. Biol.* **146**, 469 (1981).
[10] J. E. Cronan, Jr., *J. Biol. Chem.* **263**, 10332 (1988).
[11] M. D. Lane, K. L. Rominger, D. L. Young, and F. Lynen, *J. Biol. Chem.* **239**, 2865 (1964).
[12] O. Prakash and M. A. Eisenberg, *Proc. Natl. Acad. Sci. USA* **76**, 5592 (1979).

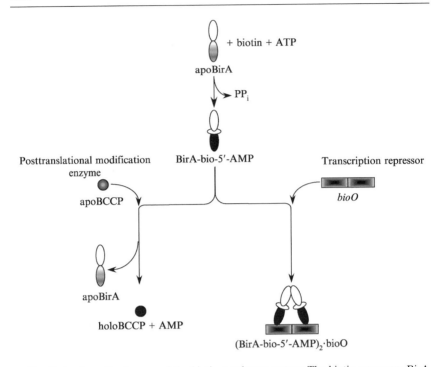

FIG. 2. A schematic diagram of the biotin regulatory system: The biotin repressor, BirA, binds to both ATP and biotin to catalyze synthesis of bio-5′-AMP. The resulting holoBirA complex either binds to apoBCCP, the apobiotin carboxyl carrier protein, and catalyzes transfer of biotin to a unique lysine residue, or binds to the biotin operator sequence to repress transcription initiation at the biotin biosynthetic operon.

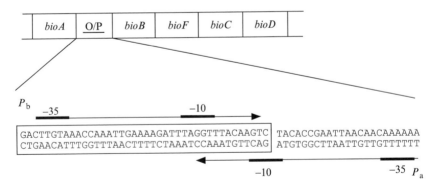

FIG. 3. A schematic of the biotin biosynthetic operon, in which the genes *bioA–bioF* encode biotin biosynthetic enzymes. Transcription is initiated from the two overlapping, divergent promoters, P_a and P_b, and binding of the biotin repressor, holoBirA, to the 40-base pair boxed sequence results in coordinate repression of transcription initiation from both promoters.

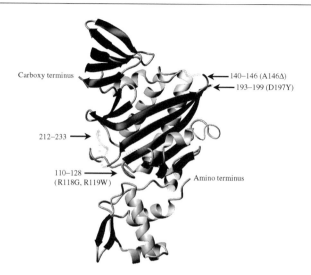

FIG. 4. A model of the three-dimensional structure of the unliganded biotin repressor, apoBirA. The springlike segments indicated by the arrows represent regions of the polypeptide where no electron density is observed and are, thus, assumed to be disordered. The model was created in MolMol,[13] using the pdb file 1BIB as input. The positions and identities of mutations that result in a loss of repression function *in vivo* are shown in parentheses.

in two previous review chapters.[14,15] The focus of this chapter is on methods to examine the coupled reactions in the biotin repressor.

Before describing developments in examining linked equilibria in the biotin repressor, some of the basic features of the system are reviewed. A model of the three-dimensional structure of the unliganded or aporepressor is shown in Fig. 4.[16] In this state the protein is monomeric and is composed of three domains: the N-terminal DNA-binding domain, the central domain, and the C-terminal domain. In addition to its structured core the central domain is characterized by four surface loops that are partially disordered in the aporepressor (Fig. 4). Binding of biotin and bio-5'-AMP to the biotin repressor (BirA) are characterized by equilibrium constants of approximately 4×10^{-8} and 4×10^{-11} M,[17] respectively, in the standard buffer conditions utilized in our laboratory. Furthermore, kinetic analysis of both binding processes by stopped-flow fluorescence reveals

[13] R. Koradi, M. Billeter, and K. Wüthrich, *J. Mol. Graphics* **14,** 51 (1996).

[14] Y. Xu and D. Beckett, *Methods Enzymol.* **279,** 405 (1997).

[15] D. Beckett, *Methods Enzymol.* **295,** 424 (1998).

[16] K. P. Wilson, L. M. Shewchuk, R. G. Brennan, A. J. Otsuka, and B. W. Matthews, *Proc. Natl. Acad. Sci. USA* **89,** 9257 (1992).

[17] Y. Xu, E. Nenortas, and D. Beckett, *Biochemistry* **34,** 16624 (1995).

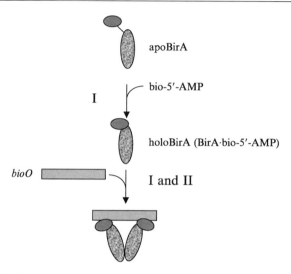

FIG. 5. Linked equilibria in biotin repressor function. (I) Binding of bio-5'-AMP is coupled to structural/dynamic changes in the protein. (II and III) Protein assembly is linked to bio-5'-AMP and DNA binding.

that binding is a two-step process, involving initial formation of a collision complex followed by a conformational change in the protein[17] (Fig. 5). The target site for biotin repressor binding, the biotin operator, is a 40-base pair inverted imperfect palindrome (Fig. 3).[18] Results of equilibrium binding measurements indicate that binding stoichiometry is two monomers per operator site and that the process is well described by a cooperative model.[19] Finally, using sedimentation equilibrium measurements, binding of the allosteric effector, bio-5'-AMP, to apoBirA has been shown to significantly enhance dimerization energetics of the protein.[20] Under standard buffer conditions the aporepressor exhibits no tendency to dimerize whereas holoBirA dimerizes with an equilibrium dissociation constant in the 10 μM range of monomer concentration. The studies described in this chapter were designed to investigate the coupled equilibria that contribute to regulation of biotin repressor function. First, results of studies of the structural/ dynamic consequences of corepressor binding for the biotin repressor monomer are described. Second, investigations of the thermodynamic properties of single-site mutants and their relationship to the results of structural probing and discussed. Third, results of high-resolution structural studies of

[18] A. Otsuka and J. Abelson, *Nature* **276,** 689 (1978).
[19] J. Abbott and D. Beckett, *Biochemistry* **32,** 9649 (1993).
[20] E. Eisenstein and D. Beckett, *Biochemistry* **38,** 13077 (1999).

the liganded protein and the relationship of the structure to the results of thermodynamic probing of the system are discussed. Fourth, investigations of the quantitative relationship between enhanced self-assembly energetics and enhanced biotin operator binding associated with effector, bio-5'-AMP, binding are presented. Finally, measurements the kinetic mechanism of biotin operator binding by holoBirA and its relationship to the functional switch from enzyme to transcription repressor are discussed.

Probing Consequences of Ligand Binding for Conformation/Dynamics of Biotin Repressor

In examining the mechanism of allostery in any protein it is critical to probe the structural and dynamic features of the system and the response of these features to binding of an allosteric effector. Results of such probing are particularly powerful for identifying regions of the protein that are significant in transmission of allosteric response. Although X-ray crystallography is indispensible for characterizing end states of the assembly process, in elucidation of the allosteric mechanism operative in any proteins it is important to characterize "intermediates" in the process. In the case of the biotin repressor the critical intermediate is the liganded form of the monomeric protein (Fig. 2). Given the properties of the repressor protein neither X-ray crystallography nor nuclear magnetic resonance (NMR) spectroscopy is a suitable tool for study of this species. In principle, several techniques have provided useful solution probes of ligand-induced changes in a protein including partial proteolysis,[21] hydrogen–deuterium exchange coupled to mass spectrometric[22] or Fourier transform-infrared spectroscopic[23] detection, or site-directed spin labeling coupled to electron paramagnetic resonance (EPR) spectroscopy.[24] In the biotin repressor system susceptibility to hydroxyl radical cleavage was initially chosen as a probe of ligand-induced changes in the protein monomer. The advantage of this method over most of those noted above includes its lack of reliance on sophisticated instrumentation and its relatively high resolution. The methodology was first developed in the Heyduk laboratory and the description provided below is a variation on the original protocols.[25]

Probing of the aporepressor and two liganded species of the protein, BirA·biotin and BirA·bio-5'-AMP, has provided information about

---

[21] M. A. Shea, B. R. Sorensen, S. Pedigo, and A. S. Verhoeven, *Methods Enzymol.* **323**, 254 (2000).

[22] A. N. Hoofnagle and K. A. Resing, *Annu. Rev. Biophys. Biomol. Struct.* **32**, 185 (2003).

[23] A. Dong, J. M. Malecki, L. Lee, J. F. Carpenter, and J. C. Lee, *Biochemistry* **41**, 6660 (2002).

[24] J. H. Zhang, G. Xiao, R. P. Gunsalus, and W. L. Hubbell, *Biochemistry* **42**, 2552 (2003).

[25] N. Baichoo and T. Heyduk, *J. Mol. Biol.* **290**, 37 (1999).

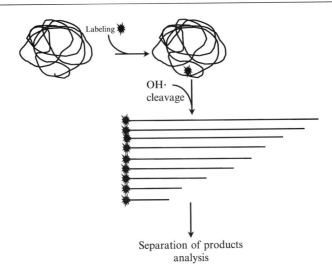

Separation of products
analysis

FIG. 6. Schematic outline of the procedure utilized for protein footprinting. The protein is first terminally labeled, either at the amino or carboxy terminus. The labeled protein is then subjected to partial chemical cleavage with hydroxyl radicals generated by Fenton chemistry. The resulting nested digestion products are separated by denaturing polyacrylamide gel electrophoresis.

structural loci of the allosteric transition in the repressor.[26] Although the substrate biotin is not the physiological corepressor for BirA it has been shown to be a weak allosteric activator.[12,27] Structural characterization of the three species was performed using hydroxyl radical cleavage of the polypeptide backbone. As indicated above, the great advantage of this technique for the biotin repressor system is that it could be performed in solution with the monomeric liganded species. Comparison of the susceptibility of the amide backbone to cleavage in the absence and presence of ligand provided information about the regions of the protein that are either directly involved in ligand binding or are conformationally/dynamically altered in response to ligand binding. The measurements are performed as follows (Fig. 6).

Production of Fusion Proteins for Labeling

A version of the BirA coding sequence in which a sequence encoding a kinase recognition and phosphorylation site, in this case VRRAS, is fused to either the amino or carboxy terminus is constructed by standard oligonucleotide directed mutagenesis methods. The fusion proteins are

[26] E. D. Streaker and D. Beckett, *J. Mol. Biol.* **292,** 619 (1999).
[27] E. D. Streaker, A. Gupta, and D. Beckett, *Biochemistry* **41,** 14263 (2002).

purified by modifications of the standard protocol[18,26] and compared with the wild-type protein with respect to ligand binding, enzymatic, and DNA-binding activities.

Phosphorylation Reactions

The cAMP-dependent protein kinase used for labeling the fusion protein is obtained from Promega (Madison, WI). Protein is labeled with ^{33}P, using [γ-^{33}P]-ATP as a substrate in the kinase reaction. Details of the reaction conditions and the purification of the labeled protein are provided in Streaker and Beckett.[26]

Hydroxyl Radical Cleavage Reactions

Solutions of the labeled protein, either ligand free or ligand bound, are prepared. In this system the equilibrium dissociation constant for binding of the ligands is low or binding is tight. It was therefore straight forward to attain stoichiometric binding. All reactions are performed in buffer containing 10 mM morpholinepropanesulfonic acid (MOPS; pH 7.5 at 20°), 200 mM KCl, 2.5 mM MgCl$_2$, and 2.0 mM CaCl$_2$. Reactions are initiated by first placing 2 μl of each of the three cleavage reagents [10 mM H$_2$O$_2$, 800 mM sodium ascorbate, and 40 mM Fe(NH$_4$)$_2$(SO$_4$)$_2$, 80 mM EDTA] on the inside wall of the reaction tube followed by mixing by gentle vortexing. The reactions are allowed to proceed for 1 h at 20°. This reaction time has been optimized to obtain sufficient cleavage of the protein as judged by phosphorimages of the gels, while leaving the majority of the protein uncut. The samples are quenched by addition of low molecular mass sodium dodecyl sulfate–polyacrylamide gel electrophoresis (SDS–PAGE) sample buffer[28] and stored at −70° until electrophoresis.

Separation of Cleavage Products

Hydroxyl radical cleavage products are separated by SDS–PAGE on a Tris-Tricine polyacrylamide gel system[28] consisting of a 16.8% T, 6% C separating gel, a 10% T, 6% C spacer gel, and a 4% T, 6% C stacking gel. T and C refer to the acrylamide monomer and cross-linker, respectively. The electrophoresis running buffer contains 0.1 M Tris, 0.1 M Tricine, and 0.1% SDS. Samples are loaded into lanes and electrophoresed through the stacking gel at 100 V and at 600 V for the remainder of the run. Gels are fixed by soaking in a solution containing 10% acetic acid and 40% methanol in H$_2$O, transfered onto 3MM Chr (Whatman, Clifton, NJ) chromatography paper, dried and exposed overnight to a storage phosphor screen (Amersham Biosciences, Piscataway, NJ).

[28] H. Schagger and G. von Jagow, *Anal. Biochem.* **166,** 368 (1987).

Data Analysis

In analysis of the footprinting data two important issues must be considered. First, accurate molecular weight markers must be run on all gels in order to allow accurate assignment of changes of reactivity in the protein sequence. These markers should be generated from the protein that is the subject of footprinting analysis. The markers generated for the biotin repressor are obtained by treatment of the protein with BNPS-skatole [3-bromo-3-methyl-2-(2-mitrophenylmercapta)-3*H*-indole], which cleaves at tryptophan residues.[29] The second consideration is that careful quantitation of the gels must be performed. In this work the gels are subjected to phosphorimaging. Line scans of the images are made and the intensities of the scans are normalized using the measured intensity of the peaks corresponding to the undigested protein. This is justified by the fact that conditions used for digestion are mild enough that >98% of the protein was intact. Difference scans for the unliganded and liganded protein are prepared by subtracting the normalized scan obtained for the latter from that of the former.

Results

The image of a gel obtained from hydroxyl radical cleavage studies of the biotin repressor is shown in Fig. 7. Visual inspection of the gel reveals that binding of the small ligands to the protein results in protection of some sites from cleavage. Line scans and difference scans obtained for products of digestion in the absence and presence of bio-5'-AMP are shown in Fig. 7, along with a map of locations of changes on the three-dimensional structure of BirA. The major conclusion of these studies is that allosteric effector binding results in significant protection from chemical cleavage of the flexible loops identified in high-resolution X-ray crystallographic studies of the aporepressor. This was true not only for the physiological effector, bio-5'-AMP, but also for the weaker effector, biotin.

Thermodynamic Probing of Single-Site Mutants: Correlation between Loss of Dimerization Function and Reduced DNA-Binding Affinity

Results of protein footprinting studies indicate the involvement of the surface flexible loops of BirA in the allosteric response. Results of *in vivo* studies of repressor mutants had indicated that mutations in these same loops result in loss of transcription repression[9] (Fig. 4). These results

[29] D. L. Crimmins, D. W. McCourt, R. S. Thoma, M. G. Scott, K. Macke, and B. D. Schwartz, *Anal. Biochem.* **187**, 27 (1990).

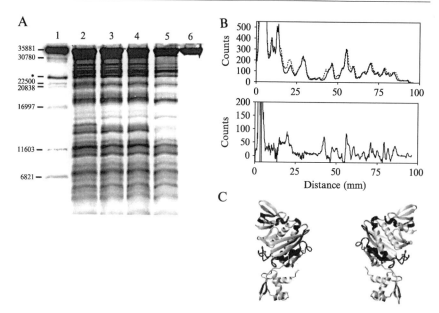

FIG. 7. Results of hydroxyl radical cleavage of the biotin repressor in different liganded states.[26] (A) Image of gel showing products of hydroxyl radical cleavage. Lane 2, apoBirA; lane 3, BirA·bio-5'-AMP; lane 4, BirA.biotin; lane 5: holoBirA·*bioO* DNA. The samples in lanes 1 and 6 are the standards generated from BNPS-skatole cleavage of BirA at tryptophan residues and the uncut protein, respectively. (B) *Top:* Line scans of data from lane 2 (dotted line) and lane 3 (solid line). *Bottom:* Difference between the two line scans shown in the upper panel. (C) Locations of protection of BirA from OH· cleavage in the three-dimensional protein structure as a consequence of bio-5'-AMP binding is shown in black. The model was generated using MolMol,[15] with pdb file 1BIB as input.

suggested that biophysical analysis of the mutants might prove fruitful in further delineating the structural basis for allostery in the protein. To this end the genes encoding the mutants were cloned and the proteins were purified and subjected to thermodynamic analysis of *bioO* binding and self-assembly.[30,31]

Subcloning, Overexpression, and Purification

Genes encoding the mutant proteins are amplified from the *E. coli* genome, using the polymerase chain reaction, and the resulting products are inserted into the expression vector pBtac2. High-level transcription of

[30] K. H. Kwon and D. Beckett, *Protein Sci.* **9,** 1530 (2000).
[31] K. H. Kwon, E. D. Streaker, S. Ruparelia, and D. Beckett, *J. Mol. Biol.* **305,** 821 (2000).

the gene from these constructs is initiated from a *tac* promoter on addition of the inducer isopropyl-β-D-thiogalactopyranoside (IPTG) to cultures grown to logarithmic phase. Each protein is purified using variations on the standard protocol.[18,30,31] The activity of each purified protein is determined by stoichiometric titration with bio-5'-AMP.

DNase I Footprinting

Measurements of binding of the mutant proteins to the biotin operator are performed by the quantitative DNase I footprinting technique. This method has been the subject of a previous review in this series[32] and the specifics of its application to the biotin repressor–biotin operator interaction are provided in Abbott and Beckett.[19] A footprint obtained from measurements of binding of the A146Δ mutant to *bioO* is shown in Fig. 8A. As indicated in the introduction to this chapter, the binding of holoBirA to the two half-sites of *bioO* can be successfully modeled by a two-site cooperative model (Fig. 8). Equations relating the fractional occupancy of each half-site to holorepressor monomer concentration are as follows.

$$Y_1 = \frac{k_1[P] + k_1 k_2 k_{12}[P]^2}{1 + (k_1 + k_2)[P] + k_1 k_2 k_{12}[P]^2} \tag{1}$$

$$Y_2 = \frac{k_2[P] + k_1 k_2 k_{12}[P]^2}{1 + (k_1 + k_2)[P] + k_1 k_2 k_{12}[P]^2} \tag{2}$$

where the fractional occupancy depends on the microscopic equilibrium constants for the binding interaction (Fig. 8B) as indicated. Binding data obtained by analysis of the footprint in Fig. 8A are shown in Fig. 8C along with best-fit curves obtained from nonlinear least-squares analysis of the data, using Eqs. (1) and (2).

Assembly Properties of Mutants Obtained by Sedimentation Equilibrium Analytical Ultracentrifugation

The monomer–dimer assembly reaction of the biotin repressor is ideally suited to measurement by equilibrium sedimentation. The mutant repressor proteins saturated with bio-5'-AMP were subjected to sedimentation analysis. Measurements are performed at multiple loading concentrations and multiple rotor speeds.[33,34] Results of global analysis of the

[32] M. Brenowitz, D. F. Senear, M. A. Shea, and G. K. Ackers, *Methods Enzymol.* **130,** 133 (1986).

[33] D. E. Roark, *Biophys. Chem.* **5,** 185 (1976).

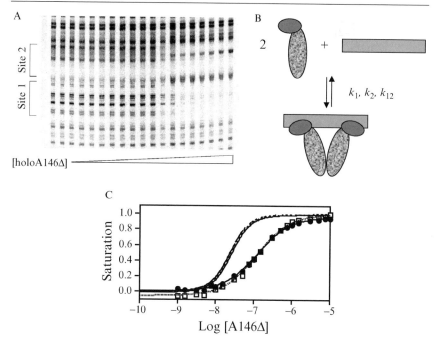

FIG. 8. DNase I footprinting analysis of the interaction of BirA mutants with *bioO*.[31] (A) Footprint obtained from titration of *bioO* with the BirA mutant A146Δ. The biotin operator DNA is titrated with increasing concentrations of the mutant protein that is saturated with bio-5′-AMP and occupancy of the operator site is probed with DNase I. The two half-sites of the biotin operator are designated sites 1 and 2. (B) Cooperative model for binding of the two repressor monomers to the two operator half-sites. The microscopic parameters, k_1 and k_2, represent the intrinsic equilibrium constants governing binding of a monomer to each half-site. The parameter k_{12} is the cooperativity constant. (C) Isotherms generated for binding of the two monomers to site 1 (●) and site 2 (□). Solid lines indicate the positions of isotherms obtained with the wild-type repressor. Lines through the data points indicate the curves simulated using the parameters obtained by nonlinear least-squares analysis of the footprinting data, using Eqs. (1) and (2).

data reveal that each mutant is well described as a single species with a molecular weight consistent with that of the monomeric protein.[35] Thus no evidence of self-assembly of any of the mutants is provided by the data.

---

[34] T. M. Laue, *Methods Enzymol.* **259**, 427 (1995).

[35] M. L. Johnson, J. J. Correia, D. A. Yphantis, and H. R. Halvorson, *Biophys. J.* **36**, 575 (1981).

TABLE I
ENERGETICS OF BINDING OF BirA VARIANTS TO BIOTIN OPERATOR[a]

	$\Delta G_{12}^{\circ\,b}$ (kcal/mol)	$\Delta G_1^{\circ\,b}$ (kcal/mol)	$\Delta G_2^{\circ\,b}$ (kcal/mol)	$(var)^{1/2\,c}$	$\Delta\Delta G_{TOT}^{\circ}$ (kcal/mol)
		200 mM KCl			
Wild type	$(-2)^d$	-9.3 ± 0.4	-9.2 ± 0.4	0.059	—
G115S	(-2)	-9.0 ± 0.4	-9.1 ± 0.4	0.061	0.4
R118G	(-2)	-7.5 ± 0.4	-7.3 ± 0.4	0.054	3.6
A146Δ	-0.2 ± 0.2	-9.2 ± 0.1	-9.2 ± 0.2	0.017	1.9
		50 mM KCl			
Wild type	(-2)	-10.8 ± 0.5	-10.3 ± 0.5	0.072	—
R119W	(-2)	-8.4 ± 0.6	-7.9 ± 0.6	0.068	4.8
A146Δ	-1.9 ± 0.5	-9.7 ± 0.2	-9.3 ± 0.3	0.034	2.2
D197Y	-1.4 ± 0.5	-8.7 ± 0.2	-8.5 ± 0.3	0.042	4.5

[a] Measured in the following buffer: 10 mM Tris (pH 7.50 at 20°), 2.5 mM MgCl$_2$, calf thymus DNA (20 μg/ml), BSA (100 μg/ml), 200 or 50 mM KCl.
[b] The free energy terms ΔG_1°, ΔG_2°, and ΔG_{12}°, are related to the microscopic equilibrium constants in Eqs. (1) and (2) by the expression $\Delta G_1^{\circ} = -RT \ln k_i$.
[c] Square root of the variance of the fit.
[d] Analyses in which the cooperative free energy is shown in parentheses were performed by fixing ΔG_{12}° at the indicated value and allowing only the intrinsic free energies, ΔG_1° and ΔG_2°, to float.

Results

The results of studies of these single-site mutants in the flexible loops indicate that they are defective in both site-specific binding to *bioO* (Table I) and self-assembly. These results are consistent with a role for the flexible loops in dimerization of BirA and with an allosteric mechanism that involves corepressor-linked dimerization of the protein.

Structural Information

The results of high-resolution structural studies have been invaluable in efforts to determine the allosteric mechanism of the biotin repressor. Crystals of the protein bound to biotin have been grown and a high-resolution structure has been determined.[36] As indicated below, biotin, although not the physiological allosteric activator for BirA, is a weak activator of binding to *bioO*. A model of the resulting structure is shown in

[36] L. Weaver, K. H. Kwon, D. Beckett, and B. W. Matthews, *Proc. Natl. Acad. Sci. USA* **98**, 6045 (2001).

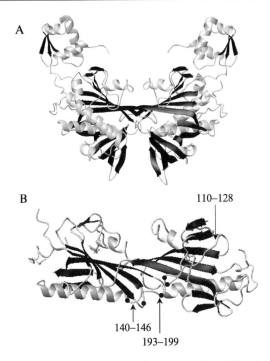

FIG. 9. (A) Model of the three-dimensional structure of the BirA–biotin dimer.[36] The model was created using MolMol[15] with the pdb file 1HXD as input. (B) Detail of the BirA homodimer interface. The biotin molecules are shown as gray stick figures under the 110–128 loops. The locations of the loops in the interface are indicated with arrows and the locations of point mutations that result in defects in repression *in vivo* and *bioO* binding and dimerization *in vitro* are represented by black circles.

Fig. 9. As indicated in the model, the biotin-bound repressor is a dimer in which the protein–protein interface is formed by side-by-side alignment of the β sheets of the central domain of each monomer. Also located in the homodimer interface are three of the four surface loops that are not visible in the structure of the aporepressor. The locations of the loops in the homodimer interface are consistent with results of hydroxyl radical probing of the allosteric switch and with results of studies on the mutants in these same loops described above. The results of structural and thermodynamic analysis suggest a mechanism of allosteric activation of BirA. The corepressor, by binding to the repressor, induces disorder-to-order transitions in the surface loops. We infer that preordering of the loops via ligand binding reduces the entropic cost of dimerization, thereby rendering the free energy of dimerization more favorable. This favorable dimerization process in turn enhances the overall stability of the transcription repressor complex.

Direct Measurements of Linkage between Dimerization and
DNA Binding

The results of combined mutagenesis and structural studies indicate
that DNA binding and dimerization are linked in the biotin repressor
system. Moreover, the results are consistent with the idea that the allosteric
effector promotes binding of the protein to DNA by promoting its self-
assembly. This hypothesized allosteric mechanism can be directly tested
by thermodynamic approaches.[27] In essence, one must determine if the
magnitudes of energetic enhancement of dimerization and site-specific
DNA binding provided by binding of the allosteric activator to BirA
are equivalent. The biotin repressor system is particularly well suited for
measurements of this type for two reasons. First, three species of the protein
including the aporepressor, the biotin-bound repressor, and the bio-
5′-AMP-bound repressor, all of which exhibit specific binding to the biotin
operator site, are available for these measurements. Second, the assembly
properties of all three species can be measured by equilibrium sedimentation. The strategy employed for quantitating the linkage between small
ligand binding, protein assembly, and DNA binding is to measure the dimerization and DNA-binding energetics of each species. These energetic parameters are then utilized to calculate the free energy change for dimerization
and for DNA binding associated with converting one species to another.
It is reasoned that if a close correspondence between these $\Delta\Delta G°$ values
exists for converting, for example, apoBirA to BirA·biotin, then one can
conclude that allosteric activation of the protein occurs solely through
enhancement of the stability of the protein–protein interaction (Fig. 10).

BioO-*Binding Properties of Three Species*

Measurements of DNA binding of the three species are made by
DNase I footprint titrations. Because both the biotin-bound and apo
species of BirA bind weakly to *bioO* these measurements are made in
buffers containing a relatively low monovalent ion concentration ([KCl]
= 50 m*M*). A potential complication associated with DNA-binding
measurements performed at low salt concentration is that nonspecific
binding of the protein to DNA will interfere with specific binding. As indicated in Fig. 11, this is not the case for apoBirA, the species characterized
by the lowest affinity for *bioO*. If nonspecific binding had been significant
one would expect to observe protection of regions outside of the operator
site in the DNase I footprints and no such protection is observed. DNase I
footprints are performed and analyzed essentially as described above
and the thermodynamic parameters governing the binding interactions
are provided in Table II.

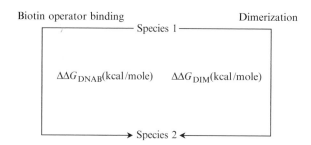

FIG. 10. Strategy used to test the thermodynamics of linkage between small ligand binding, dimerization, and *bioO* binding in the biotin repressor system. Changes in the free energies of dimerization and *bioO* binding, $\Delta\Delta G_{DIM}$ and $\Delta\Delta G_{DNAB}$, on conversion of one liganded species to another are calculated from the results of measurements of each of these processes for each pair of BirA species.

TABLE II

DNase I Footprinting Measurements of *bioO* Binding by Three BirA Species[a]

Species	ΔG_{12}[b] (kcal/mol)	ΔG_1 (kcal/mol)	ΔG_2 (kcal/mol)	$(var)^{1/2}$ [c]	ΔG_{TOT} (kcal/mol)
apoBirA	(−2)	−8.3 ± 0.5	−7.9 ± 0.5	0.048	−18.2 ± 0.7
BirA·biotin	(−2)	−9.0 ± 0.3	−8.8 ± 0.3	0.026	−19.8 ± 0.4
BirA·bio-5′-AMP	(−2)	−10.3 ± 0.3	−10.3 ± 0.3	0.03	−22.6 ± 0.4

[a] All measurements were performed in buffer of the following composition: 10 mM Tris-HCl (pH 7.50 ± 0.02 at 20°), 50 mM KCl, 2.5 mM MgCl$_2$, 1.0 mM CaCl$_2$, BSA (100 μg/ml), sonicated calf thymus DNA (20 μg/ml).
[b] In nonlinear least-squares analysis of the data using Eqs. (1) and (2), the cooperative free energy was fixed at −2 kcal/mol.
[c] Square root of the variance of the fit.

Dimerization Properties of Three BirA Species

Measurements of the dimerization of the three BirA species are performed by sedimentation equilibrium. As with any measurement of equilibrium assembly properties of a protein by this method, samples prepared at a range of initial loading concentrations and centrifuged at different speeds are employed to obtain concentration versus radial position information.[33,34] These data are subjected to global nonlinear least-squares analysis to obtain best fit parameters for the dimerization constants.[35] The data for each of the three species are well described by a monomer–dimer model and the magnitudes of the equilibrium dissociation constants for the

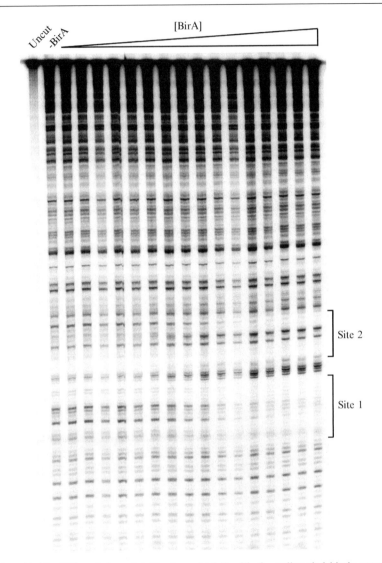

FIG. 11. The DNase I footprint titration of *bioO* with the unliganded biotin repressor apoBirA is specific.[27] The footprint was obtained in buffer containing 10 mM Tris-HCl (pH 7.50 ± 0.01), 50 mM KCl, 2.5 mM MgCl$_2$, 1 mM CaCl$_2$, sonicated calf thymus DNA (20 μg/ml), and bovine serum albumin (100 μg/ml), at 20.0 ± 0.1°.

TABLE III
RESULTS OF GLOBAL ANALYSIS OF SEDIMENTATION EQUILIBRIUM MEASUREMENTS OF
ASSEMBLY PROPERTIES OF THREE BirA SPECIES

Species[a]	Single species[b] (Da)	Monomer–dimer (K_d^b) (M)	$\Delta G^{\circ}_{DIM}{}^c$ (kcal/mol)	$(var)^{1/2\ d}$
apoBirA	39,011 (37,815, 40,447)			0.0074
		$2.0\ (1.2,\ 3.5) \times 10^{-3}$	$-3.6\ (-3.9,\ -3.3)$	0.0071
BirA·biotin	40,687 (39,490, 41,644)			0.0056
		$0.9\ (0.7,\ 1.2) \times 10^{-3}$	$-4.1\ (-4.3,\ -3.9)$	0.0049
BirA·bio- 5′-AMP	57,469 (54,454, 60,672)			0.0072
		$1.5\ (0.8,\ 2.7) \times 10^{-6}$	$-7.8\ (-7.5,\ -8.2)$	0.0071

[a] All measurements were performed in 10 mM Tris-HCl (pH 7.50 ± 0.02 at 20°), 50 mM KCl, 2.5 mM MgCl$_2$. Sufficient ligand was present in the samples containing complex to fully saturate the protein.
[b] Resolved molecular masses and equilibrium dimerization constants were obtained by global analysis of data obtained at multiple rotor speeds and loading concentrations. Loading concentrations for apoBirA and BirA·biotin were 40 and 90 μM and speeds were 20,000, 22,000, and 24,000 rpm. Loading concentrations for the BirA·bio-5′-AMP complex were 2.5, 5.0, and 7.5 μM and speeds were 22,000 and 24,000 rpm.
[c] Gibbs free energies were calculated from the equilibrium dissociation constants, using the relationship $\Delta G_{DIM} = RT \ln K_{DIM}$.
[d] Square root of the variance of the fit.

dimerization process are related in the following order: BirA·bio-5′-AMP < BirA·biotin < apoBirA[27] (Table III).

Results

Results of dimerization and DNA-binding measurements performed on the biotin repressor reveal a correlation between the increase in dimerization affinity and increased dimerization affinity. For conversion of the apo-repressor to the bio-5′-AMP-bound form, dimerization and *bioO* binding are enhanced to the same extent. However, as shown in Fig. 12, the extent to which effector binding enhances the Gibbs free energy of dimerization does not always "match" the extent to which it enhances DNA-binding affinity. In the case of the biotin-bound species it appears that there may be some uncoupling of ligand binding, dimerization, and DNA binding that warrants further investigation.

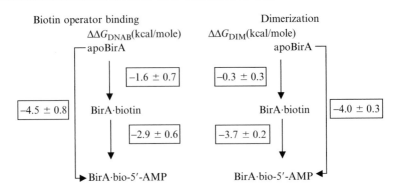

FIG. 12. Results of measurements of the energetics of dimerization and bioO binding for the three BirA species apoBirA, BirA·biotin, and BirA·bio-5'-AMP.[27] Differences in the total Gibbs free energies of bioO binding ($\Delta\Delta G_{DNAB}$) for pairs of species are shown on the left and those for dimerization energetics ($\Delta\Delta G_{DIM}$) are shown on the right.

Kinetics of Linked *bioO* binding and Dimerization of HoloBirA

Results of equilibrium binding measurements of the holoBirA–*bioO* interaction can be successfully modeled with a two-site cooperative model. However, the results of DNase I footprinting measurements can also be analyzed with a model in which a preformed dimer binds to the operator site. Ultimately, ascertaining the appropriate binding model requires kinetic measurements of the binding process. Two possible kinetic mechanisms for the holoBirA–*bioO* interaction can be envisioned (Fig. 13). In the first the two protein monomers bind in a stepwise manner to the DNA. In the second the repressor first dimerizes and then associates with the DNA. On the basis of the weak dimerization affinity of holoBirA one can formulate a persuasive argument against the preformed dimer model.[20] At the concentrations relevant to operator binding the dimer concentration is too low for it to be the relevant binding species. However, the inability to detect any operator-bound species in which a single monomer is bound suggests that the repressor dimer is the kinetically relevant species in the biotin operator-binding process.[37] The kinetic mechanism of binding was directly investigated by performing time-resolved DNase I footprinting measurements of the binding reaction.[38]

The kinetic investigations of assembly of the biotin repressor–operator complex is relevant to understanding the functional regulation of the repressor at multiple levels. First, the results of kinetic measurements

[37] E. D. Streaker and D. Beckett, *Biochemistry* **37**, 3210 (1998).
[38] E. D. Streaker and D. Beckett, *J. Mol. Biol.* **325**, 937 (2003).

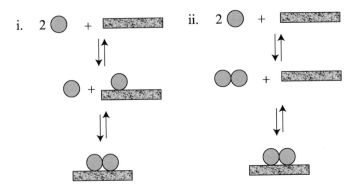

FIG. 13. Two alternative kinetic mechanisms for binding of the two holoBirA monomers to the two biotin operator half-sites: (i) stepwise binding of the two monomers and (ii) DNA binding preceded by dimerization.

provide information about the mechanism of allosteric activation of the protein. Structural probing of the protein monomer is consistent with alterations in the protein structural on corepressor binding. Thermodynamic measurements of the linkage provide information about the relationship between the enhancement in stability of the dimer relative to the monomer and its relationship to operator-binding energetics. Results of kinetic measurements alone provide information about the influence of the effector on the stepwise assembly of the repression complex. This information is not only important for determining mechanism, but is also significant for elucidation of control of partitioning of the protein between its enzymatic and site-specific DNA-binding functions. As indicated in Fig. 2, BirA functions both as a biotin holoenzyme ligase and a transcriptional repressor. Whereas the former activity is assumed to involve 1:1 heterodimerization of BirA with BCCP, in the latter a holorepressor dimer is complexed with *bioO*.[39] Elucidation of the details of regulation of kinetic partitioning between the two different functions depends on knowledge of the detailed kinetic mechanism of site-specific DNA binding.

Time-Resolved DNase I Footprinting

   Time-resolved DNase I footprinting refers to transfer of the DNase I footprint technique to the time domain. This is accomplished with rapid mixing quench-flow instrumentation. The basic outline of the measurement is that protein and DNA are rapidly mixed and the DNA is probed, using DNase I, as a function of time for occupancy of the *bioO* site. The technique

---

[39] L. Weaver, K. H. Kwon, D. Beckett, and B. W. Matthews, *Protein Sci.* **10,** 2618 (2001).

originated in the Brenowitz laboratory, where both DNase I and hydroxyl radicals have been used to measure the time dependence of protection of a site on DNA by a binding protein.[40,41] Although some minor alterations in the technique have been made in investigating holoBirA–*bioO* binding kinetics, the major implementation is the application of global nonlinear least-squares analysis for elucidation of the kinetic mechanism and the kinetic parameters associated with the binding process.

The materials utilized in time-resolved DNAse I footprinting measurements are prepared as described for equilibrium footprinting, with the exception that in order to obtain reasonable signal-to-noise ratios the amount of radiolabeled DNA utilized per reaction is considerably greater than that used for an equilibrium measurement. This higher concentration is still, however, 100-fold less than the midpoint for the protein–DNA interaction. In designing the kinetic measurements final protein concentrations over as broad a range as possible are employed. At the upper end of the concentration range this is limited by the time resolution of the quench-flow instrument used for the measurements. At the low end of the concentration range the decrease in signal to noise or the amplitude of the progress curve is the limiting factor.

A schematic diagram illustrating the experimental design utilized in the measurements is shown in Fig. 14. Protein and DNA are loaded into the two sample loops indicated. The reaction time for a particular measurement programmed into the instrument's microprocessor indicates which reaction loop is used for the mixing of the components. The reaction is initiated when the buffer from syringes A and B is pushed into the lines, thus forcing the protein and DNA into the reaction loop. The reaction is then probed with DNase I, which is added from the syringe indicated in Fig. 14. The amount of time between mixing and probing is determined by the reaction loop selected for the particular measurement. Finally, DNase I cleavage is stopped by expelling the reaction mixture into a tube placed in the exit line that contains an EDTA solution. The quenched reactions are treated like those generated in an equilibrium DNAse I footprint (Fig. 14).

In performing these footprinting measurements there are two critical experimental details. First, to avoid spillover from one time point to the next, the lines in the quench-flow instrument should be scrupulously cleaned between measurements. Instructions for doing so are included in Hsieh and Brenowitz.[40] Second, to obtain detailed information about the

[40] M. Hsieh and M. Brenowitz, *Methods Enzymol.* **274,** 478 (1996).
[41] G. M. Dhavan, D. M. Crothers, M. R. Chance, and M. Brenowitz, *J. Mol. Biol.* **315,** 1027 (2002).

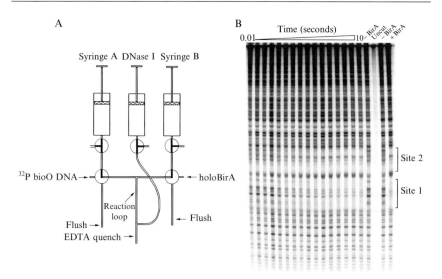

FIG. 14. (A) Diagram of the quench-flow instrument with the experimental design for measurements of time-resolved footprints of the association of holoBirA with *bioO*.[38] See text for additional explanation of the experimental protocol. (B) A time-resolved footprint measured at [holoBirA](monomer) = 4×10^{-7} *M*. Bands corresponding to the two operator half-sites are indicated.

kinetic mechanism of binding and the parameters governing the binding process it is important to perform measurements over as broad a range of protein concentrations as is experimentally feasible.

Data Analysis

The analysis of time-resolved footprinting measurements can be divided into three steps. First the images of gels are blocked to obtain the fractional occupancy versus time information at each protein concentration. These are essentially progress curves. In the second step each progress curve is subjected to nonlinear least-squares analysis to obtain time constants and amplitudes of the phases associated with the progress curve. In the final stage of analysis the data obtained at all protein concentrations are globally analyzed to ascertain which model accurately describes the data and the rate constant governing each step in the binding reaction.

Generating Progress Curves. The blocking of data to obtain fractional saturation versus time information is performed as described for equilibrium quantitative DNase I footprints.[32] The data are first represented as percent protection and are then normalized for fractional saturation to a scale from zero to one (Fig. 15).

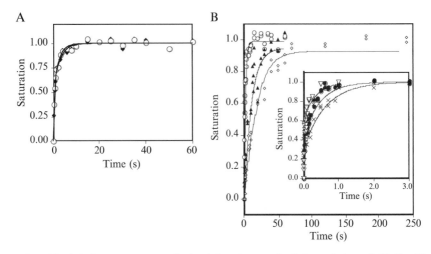

Fig. 15. (A) Progress curves obtained for time-resolved footprinting of *bioO* with holoBirA; [holoBirA](monomer) = 4×10^{-7} M.[38] Symbols indicate the individual half-site footprints and lines represent the best-fit curves obtained from nonlinear least-squares fitting of the data, using a double-exponential model. (B) Results of global analysis of kinetic measurements of the holoBirA–*bioO* interaction, using the model in which the preformed dimer binds to *bioO* [Eq. (4)]. Sixteen data sets obtained at 8 protein concentrations were used as input in the global analysis.

Initial Analysis of Progress Curves. The progress curves are subjected to nonlinear least-squares analysis to obtain information about the appropriate model for describing the progress curves and the amplitudes and time constant of each kinetic phase. The program Origin (OriginLab, Northampton, MA) is used for this step in the analysis of progress curves obtained by kinetic measurements of the biotin repressor–operator interaction. In analysis of the data, each progress curve is well described by a double-exponential model (Fig. 15). One of the phases corresponds to a burst at high protein concentrations whereas the second is slow at these same concentrations. As the repressor concentration is decreased the apparent rate of each phase decreases.

Global Analysis of the Progress Curves. The goal of kinetic measurements is to ascertain the kinetic mechanism that best describes a process and to obtain estimates of the magnitudes of rate constants associated with each step in the mechanism. This is best approached using a combination of simulation and global nonlinear least-squares analysis of progress curves obtained over a range of concentrations. Programs developed specifically for this type of analysis, KINSIM and FITSIM, were used in analysis of the time-resolved DNase I footprinting measurements performed on the holoBirA–*bioO* binding interaction.[42,43] The analysis is carried out in two stages. First, progress curves are simulated, using the different kinetic mechanisms under consideration. In this case the two models include (1) stepwise binding of two holoBirA monomers and (2) binding of a preformed dimer to the DNA. The chemical equations for these two models along with the relevant microscopic rate constants are as follows:

$$2\text{holoBirA} + bioO \underset{k_{-1}}{\overset{k_1}{\rightleftharpoons}} \text{holoBirA–}bioO + \text{holoBirA} \underset{k_{-2}}{\overset{k_2}{\rightleftharpoons}} \text{holoBirA}_2\text{–}bioO \tag{3}$$

$$\text{holoBirA}_2 + bioO \underset{k_{-1}}{\overset{k_1}{\rightleftharpoons}} \text{holoBirA}_2\text{–}bioO \underset{k_{-2}}{\overset{k_2}{\rightleftharpoons}} \text{*holoBirA}_2\text{–}bioO \tag{4}$$

The second step in model 2 is justified by the observation of the two phases in nonlinear least-squares analysis of the progress curves. Results of initial simulations of progress curves using the first stepwise model indicated that it could be eliminated from further consideration. This is because progress curves using physically meaningful rate constants exhibited no similarity to experimentally measured curves. Simulations based on the second model yielded progress curves that were in good agreement with experimental results. In performing the simulations it was important to cover the entire range of protein concentrations utilized in experimental measurements. This is because in testing a model through simulation it is important to determine whether progress curves in reasonable agreement with results of experimental measurements can indeed be obtained over the entire relevant concentration range. Once the simulations were completed global analysis of the data was performed. Aside from providing information to enable discarding of particular mechanisms, an advantage of performing the simulations is that they provided reasonable initial guesses about the magnitudes of the microscopic rate constants used in global analysis of the data (Fig. 15). Global analysis of the time-resolved footprinting data yielded the rate constants shown in Table IV. As in any kinetic analysis the validity of the parameters should, if possible, be tested against results of other independent measurements. In the case of the holoBirA–*bioO* interaction the overall dissociation rate of the complex as well as the equilibrium dissociation constant for the binding process had been independently measured. Therefore, the resolved microscopic rate constants obtained from the kinetic analysis were used to calculate the equilibrium constant and unimolecular dissociation rate. The resulting agreement

[42] B. A. Barshop, R. F. Wrenn, and C. Frieden, *Anal. Biochem.* **130,** 134 (1983).
[43] C. T. Zimmerle and C. Frieden, *Biochem. J.* **258,** 381 (1989).

TABLE IV
RESOLVED MICROSCOPIC RATE CONSTANTS GOVERNING
BINDING OF holoBirA DIMER TO $bioO^a$

Rate parameter	Value
k_1	$(5 \pm 1) \times 10^8 \ M^{-1} \ s^{-1}$
k_1	$19 \pm 9 \ s^{-1}$
k_2	$5 \pm 1 \ s^{-1}$
k_{-2}	$(7 \pm 4) \times 10^{-3} \ s^{-1}$

[a] The resolved rate parameters were obtained from global analysis of 16 data sets for progress curves measured at 8 holoBirA monomer concentrations ranging from 8.5×10^{-8} to $1 \times 10^{-6} \ M$. The Marquardt algorithm was used in FITSIM.[42,43] The 95% confidence intervals are indicated along with the resolved parameters.

obtained between the calculated and measured parameters provided increased support for the results of analysis of the kinetic data.

Summary

Combined structural, thermodynamic, and kinetic analysis of the biotin repressor has proved powerful in obtaining information about the linked equilibria in this system. This linkage is key to regulation of function in the complex biotin regulatory system. Results of qualitative analysis of many biological regulatory processes indicate that multiple linkages are common in modulating the function of participating macromolecules. Application of the approaches utilized in studies of the biotin repressor system to these other systems will yield information that will raise our understanding of the biology to a quantitative level.

Acknowledgments

The work described in this chapter was supported by NIH Grants RO1 GM46511 and S10-RR15899 to D.B.

[13] Distance Parameters Derived from Time-Resolved
Förster Resonance Energy Transfer Measurements and
Their Use in Structural Interpretations of Thermodynamic
Quantities Associated with Protein–DNA Interactions

By LAWRENCE J. PARKHURST

Introduction

Parameters that define a distance distribution for Förster resonance energy transfer (FRET) donor and acceptor pairs can be extracted from time-resolved FRET measurements. The ultimate interest in extracting such parameters is in determining distances that are relatively static on a nonosecond time scale. In turn, these distances provide information about multiple conformations of macromolecules and about the kinetics and energetics that link such conformations. Not only the mean distance, but parameters from higher moments of the distribution, in particular the width, provide information about the flexibility of macromolecules. These distances, and particularly the changes in distances, can be determined to high precision, allowing FRET to play a key role in providing highly precise distances in solution. This chapter reviews a few essentials of FRET with attention to experimental and computational methods and shows the relationship of distance parameters to macromolecular changes in the field of DNA and DNA–TATA-binding protein (TBP) interactions. These structural changes in turn are intimately linked to various thermodynamic properties.

Distribution of κ^2 and Distances

In general, FRET pairs are linked to macromolecules through flexible tethers of varying length. The flexibility of the tethers with regard to rotational motion of the transition moments of the chromophores reduces uncertainty concerning the average value of κ^2 to be used in expressing the rate constant for energy transfer, k_t. This motion can be assessed from combined measurements of time-resolved anisotropy $r(t)$ and steady state anisotropy r_{ss} to provide a cone angle for wobbling of the transition moment.[1] The calculated cone angle is sensitive to the fraction of slow phase (reflecting global motion) in the $r(t)$ decay. That fraction approaches

[1] L. J. Parkhurst, K. M. Parkhurst, R. Powell, J. Wu, and S. Williams, *Biopolymers (Nucleic Acid Sciences)* **61,** 180 (2002).

zero for large cone angles. Cofitting of $r(t)$ and r_{ss} greatly improves determination of this slow phase.[1,2] In turn, for any mutual orientation of the donor and acceptor transition moments, such a cone angle allows one to calculate an average value for κ^2 that is valid for wobbling dynamics fast on the time scale of energy transfer. As is well known, the limits of κ^2 extend from 0 to 4, with an infinity in the probability density function (PDF) at $\kappa^2 = 0$.[1,3] The PDF changes markedly, however, with even small wobble angles of the chromophores.[1,3] The upper and lower limits of the average κ^2 for a given cone angle are defined by the two canonical transition moment orientations—side by side, perpendicular to the intermoment vector, and head to tail, aligned with the intermoment vector. These limits alone considerably reduce the uncertainty in any distances determined from time-resolved FRET (trFRET). Let Ω_1 and Ω_2 be the donor and acceptor apical semicone angles, respectively, and let θ_{1R} and θ_{2R} be angles between the cone axes and the intermoment vector $\mathbf{R}$. Let $\alpha_i = (1 - C_i)/2$ and $\beta_i = (3C_i - 1)/2$, where $C_i = [1 + \cos(\Omega_i) + \cos^2(\Omega_i)]/3$. The time-averaged value for κ^2 is given by

$$\langle \kappa^2 \rangle = 6\alpha_1\alpha_2 + \alpha_1\beta_2[1 + 3\cos^2(\theta_{2R})] + \alpha_2\beta_1[1 + 3\cos^2(\theta_{1R})] + \beta_1\beta_2[\kappa^2] \quad (1)$$

where $[\kappa^2]$ is the familiar expression: $\cos(\theta_{12}) - 3\cos(\theta_{1R})\cos(\theta_{2R})$ for the angular portion of the dipole–dipole interaction energy. Assuming that there is a distribution of such cone axes results in a new distribution of κ^2 between these limits. Under the assumption that the distribution functions for κ^2 and interdye distance are separable as a product, simulations[1,3] have shown that mean interdye distances and particularly changes in such distances are little affected by details of $P(\kappa^2)$, provided the wobble angle is on the order of 60–70°. (As the cone angle approaches 90°, the distribution collapses to a single value, $\kappa^2 = 2/3$.) The advantage gained from the narrowing of $P(\kappa^2)$, however, results in a distribution of donor–acceptor distances $P(R)$ that must be determined to obtain an average interdye distance $(\overline{R})$.

Perhaps the simplest three-dimensional model for interdye distances is provided by a "dumbbell" model, in which each dye is tethered to the end of a straight line, but has available all points within a spherical region defined by the maximum extent of each tether. The end points of the line, of length R^*, are thus the centers of spheres of radii r_1 and r_2. Assuming

[2] P. R. Hardwidge, J. Wu, S. L. Williams, K. M. Parkhurst, L. J. Parkhurst, and L. J. Maher, III, *Biochemistry* **41**, 7732 (2002).
[3] L. J. Parkhurst, *in* "Near-Infrared Applications in Biotechnology" (R. Raghavachari, ed.), p. 5. Marcel Dekker, New York, 2001.

that the spheres do not overlap, one can calculate the mean interdye distance and the standard deviation of the distribution:

$$\overline{R} = R^* + (r_1^2 + r_2^2)/(5R^*); \quad \sigma = [(r_1^2 + r_2^2)/5]^{1/2}[1 - (r_1^2 + r_2^2)/(5R^{*2})]^{1/2} \quad (2)$$

For $r_1 = r_2$, $\overline{R} = R^* + 2r^2/5R^*$, and generally, $\sigma \approx (2/5)^{1/2}r$, or 0.63 r. These results are obtained by calculating the mean square distance between two points, one in each region, by straightforward integration, and by writing $1/R(i, j)$ in terms of the familiar Legendre polynomial expansion. That expansion is then used to multiply $\overline{R}^2(i, j)$ to obtain a simple expression for $R(i, j)$ for averaging over the two spherical regions. Orthogonality relations then lead to the simple expression in Eq. 2. These results show that for sufficiently long distances (e.g., 60 Å) between the two spheres, a σ value of 5–10 Å is little affected by R^* and that $\overline{R}$ is little affected by σ. Thus, σ should not *per se* be strongly affected by changes in $\overline{R}$ associated with a conformational change. Despite the finite limits on the interdye distances, the distance distribution $P(R)$ can be well approximated by a shifted Gaussian distribution,[1,3] characterized by the usual parameters $\overline{R}$, the most probable and mean distance, and σ, the standard deviation of the distribution, the half-width at 0.6067 of the peak of the distribution. If regions of the macromolecule are excluded regions for the dyes, then the distribution assumes a skewed Gaussian shape,[1,3] which can be modeled by an expansion in Hermite polynomials[1,4,5] that multiply the Gaussian. (In principle, dyes linked to the ends of DNA, e.g., have a large region available for motion; in other cases, as for a protein, the region may more closely resemble a hemisphere.) Such an extension adds a single coefficient for each additional term, compared with the three parameters if a second Gaussian is added, and is a preferred extension if required. Some care must be used in Hermite polynomial expansions, however, because significant regions of negative probability density can easily result. From this point we assume that a shifted Gaussian can adequately reflect the true $P(R)$ and confine our attention to the extraction of the two parameters $\overline{R}$ and σ and their uses.

Time-Resolved Donor-Detected FRET

Consider first energy transfer studied from the decay of the donor in time, $F_{D(A)}(t)$, or donor-detected FRET (DoDFRET). In the simplest view, involving no reversible paths (such as delayed fluorescence arising from a triplet level close to the lowest excited singlet of the donor),

---

[4] L. J. Parkhurst and K. M. Parkhurst, *Proc. Soc. Photo-Optical Instrum. Eng.* **2137,** 475 (1994).

[5] K. M. Parkhurst and L. J. Parkhurst, *Biochemistry* **34,** 293 (1995).

the decay of the donor is simply first-order decay with the decay constant a sum of all paths depopulating the excited state. For the donor alone, the reciprocal lifetime is the sum of rate constants for fluorescence (k_F) and for all radiationless processes (k_I). The intensity of the measured fluorescence is then proportional to k_F times the concentration of the excited donor. FRET merely adds an additional rate constant, k_t, in the exponential. In general, the decay of the donor alone [FD(t)] will be multiexponential, reflecting different conformers and hence different rates of internal conversion and intersystem crossing. We assume for simplicity that these are not linked directly to the distance distribution, and that each such component has the same rate constant for transfer, k_t, linked to $P(R)$. The FRET rate constant k_t can be written as A/R^6, where "A" includes universal constants, the spectroscopic overlap integral J, the fourth power of an effective refractive index for the medium separating the dyes, and κ^2 (discussed above).[1] The term A also includes the radiative or intrinsic lifetime, t_D^0, from the Einstein coefficient,[6] which can be calculated from the absorption spectrum;[7] however, A is commonly written in terms of the quantum yield of the donor and the associated average lifetime τ_D. It is common to rewrite the expression in terms of Ro, the distance for which k_t equals the rate of decay of the donor by all other paths, which must pertain to some reference state. It is at this point that confusion has arisen. Ro is linked to a specific reference state such that the expression A/R^6 for k_t must necessarily be independent of all other decay processes for the donor excited state. That is not the case if, for instance, one writes for k_t an expression such as $(1/\tau_{Di})(Ro/R)^6$, unless one is careful to change simultaneously Ro and τ_{Di}, which has not in general been the practice. It is for that reason that we often write for k_t the expression $(1/\tau_D^*)(Ro/R)^6$, where τ_D^* and Ro are linked to the same reference state so A reflects only the nondistance terms in k_t with no linkage to various decay paths for the donor excited state. This point is raised again below when distances inferred from steady state data are discussed. With the above understanding, we write for donor decay [Eq. (3)] and donor decay in the presence of the acceptor [Eq. (4)], starting with unit donor excited state at time 0,

$$F_D(t) = \sum \alpha_{Di} \exp(-\tau_{Di}^{-1} t) \tag{3}$$

and

$$F_{D(A)}(t) = \int \sum \alpha_{Di} \exp[(-\tau_{Di}^{-1} + k_t)t] P(R) dR \tag{4}$$

[6] R. A. Alberty and R. J. Silbey, "Physical Chemistry," pp. 450–451. John Wiley & Sons, New York, 1992.
[7] S. J. Strickler and R. A. Berg, *J. Chem. Phys.* **37**, 814 (1962).

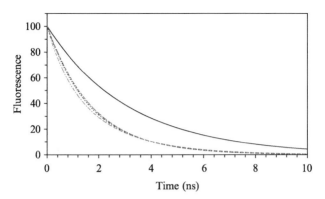

FIG. 1. Donor decay for donor alone, donor in the presence of acceptor ($\sigma = 0$), and for $\sigma = 5$ and 10 Å. The y axis is fluorescence scaled to 100 at time 0. The curves show increasingly rapid initial decay rates in the order cited above. (—) donor-only decay; (– – –) decay in the presence of the acceptor (FRET with $\sigma = 0$); (- - -) FRET with $\sigma = 5$ Å; (–··–) most rapid decay (FRET with $\sigma = 10$ Å). For these curves, Ro = 60 Å, $\tau_D = 3.2$ ns, $\overline{R} = 60$ Å, $\tau_D^* = 3.96$ ns.

where the limits of integration extend from the distance of closest approach to the maximum separation of the two dyes; in general, these limits can be set as zero and infinity with $P(R)$ understood to be normalized over this range. The normalized distribution is

$$P(R) = [1/(2\pi\sigma^2)^{(1/2)}]\exp[-R - \overline{R})^2/(2\sigma^2)] \qquad (5)$$

In the limit that σ approaches zero, $P(R)$ approaches a Dirac delta function centered at $\overline{R}$, which means that k_t reflects that single distance, and can be written $(1/\tau_D^*)(\text{Ro}/R)^6$. The effects of the $P(R)$ parameters on the donor decay are shown in Fig. 1, which shows the donor decay with no transfer, decay for transfer at a single distance, $\overline{R}$, and decay for transfer at that $\overline{R}$ with two values of σ, 5 and 10 Å. Although in principle Eqs. (4) and (5) imply an infinite number of exponentials, with τ values ranging from τ_D (for $R = $ infinity) downward to a lifetime corresponding to the distance of closest approach of the dyes, the curve for $\sigma = 10$ Å is not strikingly different from that for $\sigma = 0$. In fact, we have found in extensive simulations[8] and fitting that $P(R)$ can generally be characterized well with but two lifetimes for the pure transfer portion of the decay, or, in terms of only two corresponding distances.

[8] R. Powell, Ph.D. dissertation. University of Nebraska, 2001.

Equations (3) and (4) can be transformed directly into the frequency domain for fitting phase and modulation data.[4,5] It can be seen from Eq. (2) that the transfer process can be factored from the pure donor decay, meaning that the data $F_{D(A)}(t)$ and $F_D(t)$ can be used directly in fitting for $P(R)$ parameters without fitting for α values and τ values. In the time domain, the matter is somewhat more complex owing to convolution of the signal with the instrument response function.

The overall transfer function of an instrument is the ratio of the Laplace transform of the output signal to the Laplace transform of the input signal. Because the Laplace transform of a Dirac delta function is unity, the transfer function is simply the Laplace transform of the instrument response to a Dirac delta function, simulated by the Raleigh scattering of the incident laser pulse. Inversion of the transfer function to the time domain generates the "instrument response function" or IRF, which, when convoluted with the true input response, produces the observed signal:

$$F(t)(\text{output}) = \int \text{IRF}(t - \tau)F(\tau) \ (\text{input}) \ d\tau \tag{6}$$

with integration limits of 0 and t.

Because the Rayleigh wavelength is not that of the emitter, a correction factor (a red-shift parameter) is commonly used to account for the lower energy of the photoelectrons actually involved in generating $F(t)$ (output).[9] An alternative might be to use an appropriately weighted average of the Rayleigh and Raman scattering signals. If $F(t)$ (input) is for a fluorophore decay that is a single exponential $[a_1 \exp(-r_1t)]$, then it is easily shown[10] that the normalized IRF is given by

$$\text{IRF}(t) = (1/a_1)d[F(t) \ (\text{output})]/dt + (r_1/a_1)F(t) \ (\text{output}) \tag{7}$$

Even with digital smoothing to obtain the derivative, considerable error is introduced into IRF(t) by numerical differentiation of data. For multiple exponential decays, a simple relation analogous to Eq. (7) does not obtain. If IRF(t) is known, in principle $F(t)$ can be obtained in a model-independent fashion, for instance, by iterative deconvolution as proposed by van Cittert,[11] but with a choice other than $F(t)$ (output) for the first estimate of $F(t)$ (input). In software, such as that employed by PTI (Photon Technology, Lawrenceville, NJ) for the LaserStrobe instruments, a model, such as a sum of exponentials, is used for $F(t)$ (input),

[9] D. O'Connor and D. Phillips, "Time-Correlated Single Photon Counting." Academic Press, London, 1984.
[10] P. Wahl, J. C. Donzel, and B. Donzel, *Rev. Sci. Instrum.* **45**, 28 (1974).
[11] P. H. van Cittert, *Z. Physik.* **69**, 298 (1931).

and the parameters are adjusted in a routine that minimizes the weighted variance to best fit the observed output. This procedure is one of iterative reconvolution, which in addition to the two parameters for each exponential decay, employs one small correction for the red shift, and another small offset constant. For multiexponential decays, the optimum α values and τ values that characterize $F(t)$ (input) are obtained.

As with all minimization routines, local minima are frequently encountered, and one must refit the data numerous times to obtain a "best" fit. We find that the Durbin–Watson parameter that tests for serial correlation of the residuals is an excellent means for discriminating among fits with satisfactory values of the reduced χ^2 parameter. The original papers should be consulted for upper and lower bounds of this parameter.[12,13] In general, we find that two and at times three exponentials give excellent fits to the data. On the other hand, there are several subtle points that should be addressed. The ratio of steady state intensities for the donor in the presence of acceptor to that alone, corresponds to $\int F_{D(A)}(t)\, dt / \int F_D(t)\, dt$, where care must be taken to assure that donor concentrations are identical. That ratio places a constraint on the allowed $\{\alpha, \tau\}$ values for the corresponding time-dependent decays, and corresponds to $<\tau_{D(A)}>/<\tau_D>$, where the terms are average lifetimes for the indicated processes, and where, for instance, $<\tau_D> = \sum \alpha_{Di}\tau_{Di}$. A small amount of a fast component contributes little to the overall steady state intensity, but on normalization of the α values the effect is to suppress the remaining α values, thereby decreasing the calculated average lifetime. It is thus important to determine to what extent a given fraction of a fast lifetime can be determined. For this purpose, the IRF can be measured and fit analytically for either analytical integration of Eq. (4) or numerical integration, using the acquired discrete points. Excellent fits to the IRF can be obtained with split shifted Gaussians, one-half for times before the peak of the IRF, and one-half for times after the peak, with the latter also having a small component of exponential decay or stretched exponential to fit the long time tail of the IRF. It is then important to determine the noise profile of the decays by direct determination of the unweighted residuals from multiple runs on the same sample. A properly tuned instrument gives successive decays with no apparent correlations of successive runs, for example, no progressive decays in intensities, and no decay curve with excessive noise. A plot of the standard deviation of the residuals versus the square root of intensity should be fit well by a straight line, indicating that the main source of noise is shot noise. The slope

[12] J. Durbin and G. S. Watson, *Biometrika* **37,** 409 (1950).
[13] J. Durbin and G. S. Watson, *Biometrika* **38,** 159 (1951).

reflects the amplitude of the noise and how well the laser has been tuned and the optics aligned. One can then convolute known input functions [$F(t)$ (input)], add the appropriate Gaussian distributed random noise with σ (noise) corresponding to the instrument, and determine to what extent a fast phase can be reliably detected. For instance, a component, using the LaserStrobe, having a decay of 3 ps (which we have obtained in data fitting) is clearly spurious, and will lead, through the normalization procedure, to incorrect α values and average lifetimes. Agreement with steady state intensity ratios assures that no significant fast component is neglected. Adherence to the above-described procedures will aid in the collection of valid data for determining $P(R)$ parameters.

In measuring $P(R)$ for DNA and DNA–protein complexes, we employ a doubly labeled strand hybridized to an unlabeled strand.[4,5,14–17] When working at low concentrations (e.g., 10 nM), even well below the T_m, there will be detectable unhybridized labeled strand for 1:1 stoichiometry. For that reason, we typically employ at least a 5-fold excess of unlabeled strand, and wait 12 h for equilibrium to be reached, monitoring the emission spectra as a function of incubation time. Detectable unhybridized labeled strand will lead in general to values of σ that are too large, and with σ and $\bar{R}$ having a component that is strongly salt-dependent.[5] Studies using 1:1 stoichiometries of chains *each* singly labeled when used in DoDFRET will be particularly susceptible to these problems at low concentrations.

In principle, one could fit the actual observed $F_{D(A)}(t)$ data [Eq. (6)], using a theoretical curve similar to the right-hand side of Eq. (4), but with an outer convolution integral over the IRF. In practice, that is not what is done, provided one has a set of lifetime parameters that characterize $F_{D(A)}(t)$ and $F_D(t)$ as described above, and which have satisfied the various criteria given above as well as various criteria for goodness of fit (χ^2, runs test,[18] Durbin–Watson parameter limits, autocorrelation, and a runs analysis of the autocorrelation). The integral over $P(R)$ can then be carried out by any of various means (we employ a Simpson's rule integration, written as a weighted trapezoidal and midpoint integration, subdividing "R" until convergence is reached for all three integrations). The numerical integration is embedded in a Simplex routine, in which multiple starting points

[14] K. M. Parkhurst and L. J. Parkhurst, abstract *in* "11th Int. Congr. Photobiology." Kyoto, Japan, 1992.

[15] K. M. Parkhurst and L. J. Parkhurst, *Biochemistry* **34**, 285 (1995).

[16] K. M. Parkhurst, M. Brenowitz, and L. J. Parkhurst, *Biochemistry* **35**, 7459 (1996).

[17] K. M. Parkhurst, R. Richards, M. Brenowitz, and L. J. Parkhurst, *J. Mol. Biol.* **289**, 1327 (1999).

[18] F. S. Swed and C. Eisenhart, *Ann. Math. Stat.* **14**, 66 (1943).

are used to approach a global minimum for $\overline{R}$ and σ. Numerous data sets of pairs of $F_D(t)$ and $F_{D(A)}(t)$ are combined to obtain sets of $\{\overline{R}, \sigma\}$ values. Parameter error estimates are derived from parameters calculated from repeat runs.

Method of Moments Procedure for $P(R)$ Parameters

For DoDFRET, it is understood that one has separated spectroscopically the donor decay from that of the acceptor. There is under these conditions a procedure for determining the $P(R)$ parameters with little or no further fitting beyond extraction of the $\{\alpha, \tau\}$ parameters. Consider the function $G(t)$,[2] the ratio of $F_{D(A)}(t)/F_D(t)$, which, from Eqs. (3) and (4), is

$$G(t) = \int P(R) \exp[-(1/\tau_D^*)(Ro/R)^6] dR \qquad (8)$$

which is a function of the $P(R)$ parameters $\overline{R}$ and σ. Consider the nth moment of $G(t)$, defined as

$$\mu_G^n = \int t^n G(t)\, dt = \int \int P(R)\, \exp[-(1/\tau_D^*)(Ro/R)^6] dR\, dt \qquad (9)$$

$$\mu_G^n = \Gamma(n + 1)(\tau_D^*/Ro^6)^{n+1} \int (R^6)^{n+1} \exp[-R - \overline{R})^2/2\sigma^2](1/N)\, dR \qquad (10)$$

where $1/N = [1/(2\pi\sigma^2)^{1/2}]$, the integrations over R in Eqs. (7) and (8) are all from zero to infinity, as is the integration over time; Γ is the gamma function, which allows nonintegral moments. Provided that $P(R)$ has negligible amplitude at $R = 0$, one can transform the above integral, using the substitution $Z = (R - \overline{R})/\sigma$. With this substitution for Z, $dR = \sigma\, dZ$, the lower integration limit becomes $-\overline{R}/\sigma$, which, for $P(R)$ negligible at $R = 0$ (which is generally the case,[5] except possibly for very flexible macromolecules), allows the integral over Z to extend from $-\infty$ to $+\infty$. One then has

$$\mu_G^n = \Gamma(n + 1)(\tau_D^*\overline{R}^6/Ro^6)^{n+1} \int [wZ + 1]^{6(n+1)}(1/N) \exp(-Z^2/2)\sigma dZ \qquad (11)$$

where $\sigma/N = (2\pi)^{-1/2}$, and where $w = \sigma/\overline{R}$. One then has an integral with a closed form solution in terms of a series in even powers of w. (By symmetry, the odd powers of w vanish.) Denote the integral in Eq. (11) as S_n, corresponding to moment μ_G^n. Explicitly,

$$S_0 = 1 + 15w^2 + 45w^4 + 15w^6 \qquad (12)$$

From Eq. (11) it can be seen that the coefficients of even powers of w arise from the binomial coefficients in the expansion of $[w \cdot Z + 1]^{6(n+1)}$, multiplied by the value of the integral

$$I^{2J} = \int [wZ]^{2J}(1/N)\exp(-Z^2/2)\sigma \, dZ = 1 \cdot 3 \cdot 5 \ldots \cdot (2J-1) \qquad (13)$$

For this integral, $J = 1$ to $3(n + 1)$, for $6(n + 1)$ an even integer, and $J = 1$ to $[6(n + 1) - 1]/2$ for $6(n + 1)$ an odd integer.

Because $\Gamma(1) = \Gamma(2) = 1$, one can then see that the ratio of the square root of μ_G^1 to μ_G^0, which we denote by $\rho_{1,0}$ is

$$\rho_{1,0} = (\mu_G^1)^{1/2}/\mu_G^0 = (S_1)^{1/2}/S_0 \qquad (14)$$

which is a function only of w. Explicitly,

$$S_1 = 1 + 66w^2 + 1485w^4 + 13,860w^6 + 51,975w^8 + 62,370w^{10} + 10,395w^{12} \qquad (15)$$

Figure 2 shows $\rho_{1,0}$ as a function of w to rather large values of w, for example, $\sigma = 20$ Å for $\overline{R} = 60$ Å. In practice, a determination of $\rho_{1,0}$ allows w to be determined from such a graph, or analytically, for instance, by a Newton–Raphson procedure.

Note that Eq. (14) can be extended to values of $n < 1$, in particular, values of $n = 1/6, 1/3, 1/2, 2/3$, and $5/6$ lead to integral powers of w, in which case

$$\rho_{n,0} = (\mu_G^n)^{1/(n+1)}/\mu_G^0 \qquad (16)$$

The moment μ_G^0 is simply the area under the $G(t)$ curve or the average lifetime of the donor for only the FRET process as a decay path. This moment

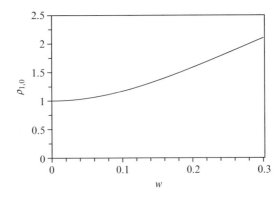

FIG. 2. The parameter $\rho_{1,0}$ as a function of w ($\sigma/\overline{R}$).

can be calculated from the $\{\alpha, \tau\}$ parameters for $F_{D(A)}(t)$ and $F_D(t)$. One constructs the two time decays, ratios the decays, and obtains a curve, which for μ_G^0 decays to zero because the added FRET path for decay of the donor excited state will assure that the numerator approaches zero more rapidly than does the denominator. In principle, this ratio function can be multiexponential, but in our experience to date the curve is well fit by only two exponentials. Only a single exponential would result if there were zero width to the distance distribution. If the $G(t)$ curve is well fit by sums of exponentials, then the moment μ_G^0 is simply $\sum \alpha_i \tau_i$, and the moment μ_G^n is $\Gamma (n + 1) \sum \alpha_i \tau_i^n$. Such a fitting can be carried out by any of a variety of minimization methods, including Simplex; alternatively, a method, such as that of Foss,[19] which eliminates a search routine, can be used. Alternatively, $G(t)$ can be constructed in a spreadsheet, and rectangular integration used to calculate any of the moments, provided that care is taken to assure that $t^n G(t)$ converges well to zero. If this is not the case, it is because the interdye distance is large and the curve $F_{D(A)}(t)$ decays only marginally more rapidly than does $F_D(t)$. In this case, moments with "n" sufficiently less than 1 may be required. For instance, for $n = 1$, the longest lifetime in a fit to $F_{D(A)}(t)$ must be less than 0.73 of the longest lifetime of $F_D(t)$ for convergence, whereas for $n = 1/2$, the comparable number is 0.90. (For the rectangular integration, the step size in time is decreased progressively until the desired convergence is reached, often in the range of 10 ps.) Thus, $\rho_{n,0}$ can be determined without actual curve fitting beyond that for determining the $\{\alpha, \tau\}$ parameters for $F_{D(A)}(t)$ and $F_D(t)$, parameters provided by commercial software. The general procedure using $\rho_{n,0}$ is essentially the same as for $\rho_{1,0}$. The procedure using $\rho_{1,0}$ is as follows: (1) from $\rho_{1,0}$, determine w; (2) knowing w, calculate S_0; (3) $\bar{R}$ is then calculated from $\bar{R} = Ro$ $(\mu_G^0/\tau_D^*)S_0^{-1/6}$ (Fig. 3 shows $S_0^{-1/6}$ as a function of w); (4) the parameter σ, equal to $w \cdot \bar{R}$, is then obtained.

Because the leading term of S_0 is 1, it is clear that when the width of $P(R)$ is near zero $\bar{R}$ is determined solely by the moment μ_G^0, which derives from a single exponential for $G(t)$. Because a biexponential $G(t)$ can well represent $P(R)$, the width of the distribution $P(R)$ must be related to the extent to which $G(t)$ differs from a single exponential. This can be assessed from $\rho_{1,0}$ in the following way. Let the longer lifetime in $G(t)$ be τ_2, and let $y = \tau_2/\tau_1$. Because $\alpha_1 = 1 - \alpha_2$, $\rho_{1,0}$ can be written in terms of α_2 and y as

$$\rho_{1,0} = [(1 - \alpha_2) + \alpha_2 y^2]^{1/2}/[(1 - \alpha_2) + \alpha_2 y] \tag{17}$$

[19] S. D. Foss, *Biometrics* **26**, 815 (1970).

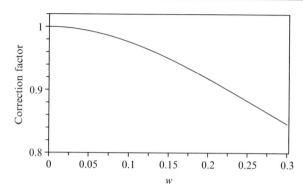

FIG. 3. The correction factor $S_0^{-1/6}$ as a function of w ($\sigma/\overline{R}$).

which shows that w, and hence σ, is determined by both α_2 and y, the latter the ratio of the two decay constants in $G(t)$ for FRET, arising from the distribution $P(R)$. $\rho_{1,0}$ does indeed increase as y increases, but $\rho_{1,0}$ converges to $(1/\alpha_2)^{1/2}$ as y approaches ∞, becoming independent of the FRET lifetime heterogeneity. That means that there are upper bounds on w ($\sigma/\overline{R}$) for every value of α_2. For α_2 decreasing from 1 to 0.1 in steps of 0.1, the limiting values of w are 0, 0.0553, 0.0827, 0.1078, 0.1339, 0.1631, 0.1989, 0.2472, 0.3250, and 0.5080.

Distances from Steady State FRET

Time-resolved FRET clearly has greater information content than steady state FRET because steady state FRET (ssFRET) can be calculated from trFRET, but not the reverse. In contrast to some intrinsic probes (Tyr, Trp, NADH, etc.), chromophores attached to a macromolecule by flexible tethers have a width for the $P(R)$ distribution that must be considered explicitly in order to extract the average interdye distance. Let I_1 be the time integral from zero to infinity of $F_D(t)$ in Eq. (3). This will be proportional to the measured steady state donor-only emission. Correspondingly, let I_2 be the integral for $F_{D(A)}(t)$ in Eq. (4), which is equal to $<\tau_{DA}>$. Assuming that the width of $P(R)$ can be explicitly neglected, then I_2, for multiexponential decay of the donor, is: $I_2 = \sum \alpha_i \tau_i/(1 + \tau_i k_t)$, where the information on an "average" distance is contained in k_t, but the individual α values, and τ values, which are unknown, are also contained in $I_1 = \sum \alpha_i \tau_i = <\tau_D>$. For biexponential donor decay, which is frequently the case, one has two equations and four unknowns. Assume next that I_1 and I_2 can actually be represented by a single donor component, in which case we can write $I_2/I_1 = (1 + I_1 k_t)^{-1}$. The ratio on the left-hand side results in cancellation of scale factors, provided identical donor concentrations are used in the two cases. This does not occur on the right-hand side, because the relation of a measured I_1 to $<\tau_D>$ is generally unknown, unless one calculates τ_D° and measures the quantum yield of the donor for the exact environment of that donor in the FRET experiment. If the donor decay is monoexponential and $<\tau_D>$ equals τ_D^* in the reference state for Ro, the term $I_1 k_t = (Ro/R)^6$. Provided the donor decay is monoexponential and that $\sigma = 0$, if $<\tau_D>$ and τ_D^* are within a factor of 1.5, the error in "R" will be under 7%. For a more realistic assessment, suppose that one simulates I_1 and I_2 for conditions where the donor concentrations for I_1 and I_2 are exactly equal, Ro = 60 Å, $\sigma = 10$ Å, $\tau_D^* = 3.96$ ns, $\overline{R} = 50$, 60, and 70 Å, and the donor has a true single decay of 3 ns. The "R" values from steady state simulations corresponding to $\overline{R}$ are, in this case, 51.0, 59.04, and 67.61 Å. If, from careful quantum yield measurements, we can equate $<\tau_D>$ and τ_D^* at 3.96 ns, we obtain "R" values of 51.4, 59.5, and 66.9 Å for the corresponding $\overline{R}$ values, a measure of the effect of the neglected σ on the estimates of "R" from steady state data alone. Compensating errors and approximations result in "R" values that differ from the $\overline{R}$ values by 0.5–3 Å, but the overall range of distances is reduced, and changes in "R" are significantly less than the changes in $\overline{R}$. Recall, however, that the above-described simulation is for a single lifetime for the donor alone and for an exact equality between $<\tau_D>$ and τ_D^*. Discrepancies between $\overline{R}$ and "R" from ssFRET for biexponential donor models have been discussed elsewhere.[3]

Acceptor-Detected FRET: ADFRET

Provided the acceptor is itself fluorescent, acceptor-detected FRET (ADFRET) provides another means for determining $\overline{R}$ and σ. There is one clear advantage of ADFRET over donor-detected FRET. The reliability of DoDFRET requires that for every donor there be an acceptor, otherwise implementation of Eq. (4) above cannot be valid. For ADFRET, however, one ideally (see discussion below) detects emission from only those acceptors having an associated donor. Suppose, for instance, that one has two reactive cysteines in a protein P. If, for instance, the maleimido esters of donor and acceptor are equally reactive toward cysteine, equimolar but excess amounts of reactive donor and acceptor probes would produce as products equal concentrations of four species: PD_2, PA_2, PDA, and PAD, the latter two differing in the specific labeling sites. Only the last two species will participate in ADFRET. Species PD_2 and PA_2 can be separately prepared and the $F_D(t)$ and $F_A(t)$ curves characterized. If, for instance, both curves are biexponential, the simplest assumption is that each of the two specific sites in P is biphasic. This hypothesis can often be tested, and corrected, by examining aliquots of P partially reacted with D and with A, because cysteines differ markedly in their reactivity.[20] Thus, if the preexponentials for 10% reacted P differ from those for PD_2 and PA_2, one can reduce the problem to the consideration of two rather than four terms in the ADFRET modeling. In such modeling, one assumes the same $P(R)$ for PDA as for PAD. One obvious advantage of ADFRET under these conditions is the fact that one need not purify the mixture. A second conceptual advantage of ADFRET is that the ADFRET directly reflects FRET, whereas for DoDFRET, FRET is detected as a difference from the donor-only decay. There are two qualifications, however. First, most acceptors have some small absorbance in the region of donor excitation and thus some of the acceptor emission arises from direct excitation that must be subtracted from the ADFRET signal. This correction becomes increasingly important as $\overline{R}$ increases. This correction can be made by measuring the decay of the acceptor-only labeled molecule at the same acceptor concentration as for ADFRET. This decay will be a decreasing exponential, whereas the true ADFRET signal begins at zero at zero time (after deconvolution of the IRF, for measurements in the time domain), rises to a maximum as the acceptor is pumped to the excited state by FRET, and then decays at a rate that approaches that for the pure acceptor. The following kinetic scheme, which assumes irreversible paths, makes the matter clear:

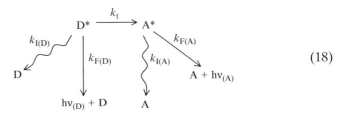

$$\tag{18}$$

Equation (18), as far as detection of ADFRET is concerned, is a simple modification of the familiar I $\rightarrow$ II $\rightarrow$ III kinetic scheme, adding an additional decay path for the donor excited state, I $\rightarrow$ D. The species III represents photons from the acceptor, and the intensity of acceptor fluorescence for ADFRET is proportional to $k_F(\text{acceptor}) \cdot II(t)$. The first step, I $\rightarrow$ II, has a rate constant k_t. The species $II(t)$ is simply $A^*(t)$. The quantity $F_{A(D)}(t)$ for the above scheme is readily derived, for instance by Laplace transforms, to give, for a single distance and single components for donor and acceptor (extensions to multiple decays is obvious and straightforward):

[20] G. Geraci and L. J. Parkhurst, *Biochemistry* **12**, 3414 (1973).

$$F_{A(D)}(t) = k_{F(A)}[D^*(0)]k_t\{\exp[-(\tau_D^{-1} + k_t)t]$$
$$- \exp[-\tau_A^{-1}t]\}/[\tau_A^{-1} - (\tau_D^{-1} + k_t)] \tag{19}$$

In practice, there is often some of the donor decay $F_{D(A)}(t)$ that passes a filter or monochromator and adds to the above-described signal. That signal will derive from the same excitation source and geometry as for $F_{A(D)}(t)$, but will be modified by the transmission characteristics of the filter/monochromator. That signal will therefore be $k_{F(D)} \exp[-(\tau_D^{-1} + k_t)t]$ multiplied by the appropriate filter transmission factor. In an instrument such as the LaserStrobe from PTI with dual detection, the correction is easy. One measures donor $F_D(t)$ on both detectors (with the same filter to remove acceptor emission) to establish the scaling of the two detectors. One then places the appropriate filter/monochromator in the second channel to be used for ADFRET to determine the magnitude of the correction to be applied to the leakage of the donor signal into the ADFRET detection channel. One then makes the appropriate measurements on the sample, measuring DoDFRET with one detector, and ADFRET with the other, arriving at a corrected ADFRET after deconvolution. Assuming that has been done, Eq. (19) must be modified by multiplication of the above expression by $P(R)$ and integration over the appropriate distance range to obtain ADFRET(t). It is clear that this case is somewhat more complex than for the donor, in that k_t appears in three places in Eq. (19), each affected by $P(R)$. ADFRET is not amenable to a methods of moments analysis. It is also clear that the information concerning $P(R)$ will appear in the rise characteristics of ADFRET rather than in the decay. Note that if there is a statically quenched form of the acceptor, which may have a different value of $\bar{R}$ from that of the fluorescent form, there can be two $P(R)$ distributions and thus two values of k_t for the donor. On the other hand, there will be only one $P(R)$ for k_t for ADFRET, that for the distance between the fluorescent form of the donor and that of the acceptor.

Before considering the consequences of $P(R)$ on ADFRET, consider Eq. (19), where we can arbitrarily set $k_F = 1$, because we wish to consider the shape of $F_{A(D)}(t)$ as a function of changes in k_t. This problem is best explored in terms of dimensionless quantities. Let $t' = k_t t$, $K = (1/\tau_A)/k_t$, and $L = (1/\tau_D)/k_t$. If we let $D^*(0) = 1$ (so "II" is scaled accordingly), one can calculate the time t'^* (and thus the actual time t^*) for which the acceptor emission reaches a maximum. The corresponding signal will be proportional to A^*. The expressions are

$$t'^* = (1/[K - (1 + L)]) \log[K/(1 + L)] \qquad t^* = t'^*/k_t \tag{20}$$

$$A^* = [1/(1 + L)][K/(1 + L)]^{[K/(1+L-K)]} \tag{21}$$

For fixed τ_D and τ_A, as k_t decreases with increasing distance R, A^* decreases and t^* increases. It is thus clear that the effect of increasing σ will be to elevate the maximum and shift the maximum to shorter times because of the biased weighting of distances shorter than $\bar{R}$. Figure 4 shows a plot of 1000 times A^* versus t^* in nanoseconds, for Ro = 60 Å, $\tau_D^* = $ 3.96 ns, $\tau_D = $ 3.2 ns, $\tau_A = $ 5 ns, for R values of 40, 45, 50, 55, 60, 65, 70, 75, and 80 Å. These parameters model ADFRET for end-labeled fluorescein–X-rhodamine oligonucleotides. It can be seen that there is nearly a linear relationship between peak height and the corresponding time. The first point that is clear is that distances can be extended to 80 Å or beyond if one raises the concentration, for instance, to be some 4-fold greater than that for conditions corresponding to $R = 60$ Å. For the same parameter values, in contrast, the average decay time for DoDFRET is decreased by just over 10% compared with that for the donor alone. Figure 5 shows a simulated decay for ADFRET for the above-cited parameters for Fig. 4 with $\bar{R} = 60$ Å and for $\sigma = 0$, and 10 Å. These curves should be compared with those for DoDFRET in Fig. 1. It should thus be possible to extract $P(R)$ parameters from ADFRET alone, but using dual detection of DoDFRET and ADFRET and global fitting should lead to better determination of these parameters, particularly when linked as well to constraints imposed by steady state measurements, as discussed above. Finally, Fig. 6 shows raw data for ADFRET for a 14-mer 5′-X-rhodamine, 3′-fluorescein-oligonucleotide as a double strand, with excitation at 488 nm and emission passing a filter (2403; Corning, Corning, NY), but before removal of the donor or acceptor emission arising from direct excitation. These corrections will yield a curve that begins at zero at time 0.

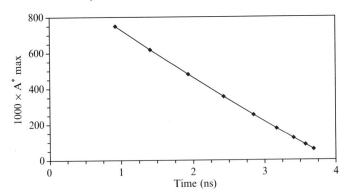

Fig. 4. One thousand times the maximum concentration of the acceptor excited state (A^*) relative to unit donor excited state plotted against time. The times for the nine maxima increase with increasing donor–acceptor separations: 40, 45, 50, 55, 60, 65, 70, 75, and 80 Å. Parameter values: Ro = 60 Å, $\tau_D = $ 3.2 ns, $\tau_A = $ 5 ns, $\tau_D^* = $ 3.96 ns.

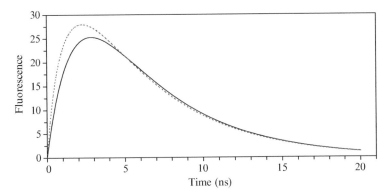

FIG. 5. Calculated acceptor-detected FRET for $\sigma = 0$ (——) and 10 Å (- - - -) for $\overline{R} = 60$ Å. Other parameters are as for Fig. 4, selected to model FRET for the donor fluorescein and the acceptor x-rhodamine. The y axis of fluorescence intensity assumes k_F values of 1, and is scaled to a value of 100 for the initial donor (D^*) concentration.

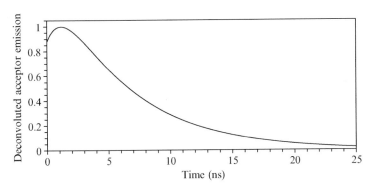

FIG. 6. Deconvoluted acceptor-detected FRET before correction for direct excitation of the acceptor X-rhodamine and correction for a small component of the donor emission. The sample was a 14-mer double-stranded AdMLP oligonucleotide with donor fluorescein and acceptor X-rhodamine. The excitation was from a nitrogen laser pumped dye laser (PTI LaserStrobe) using the dye (PTI) PL481 for excitation at 488 nm. The emission was filtered through a 659-nm cut-on Corning 2403 glass filter.

Uses of Distance Parameters

In our research with DNA oligonucleotides, we have employed fixed sequences of three or four bases at the 5' and 3' ends with a view toward obtaining nearly constant interactions of dyes with the DNA in order to assess effects of sequence and protein binding in the center of the oligonucleotides. Furthermore, we have used ends for which there is negligible change in either donor or acceptor fluorescence on hybridization of the single strand, implying, together with the time-resolved anisotropy results,[1,2,21] that there is little interaction between the dyes and the DNA, allowing them to function as probes and not as perturbants. The preceding discussion made no interpretation of the parameter σ, other than in terms of a simple model for motion of a particle on a tether at the end of a rigid rod. Such motion, represented by relatively static distributions on the nanosecond time scale, although a concomitant of the labeling chemistry for most extrinsic probes, is advantageous in providing relatively sharp distributions of κ^2, allowing $P(R)$ to be attributed largely to the tether and macromolecule motions. Separation of the dye distribution from the total distance distribution should allow insight into the multiplicity of conformations of the macromolecule and of the macromolecular dynamics.

Although DNA bending has been studied in solution[22–30] for DNA sequences far longer than can be studied by FRET using dyes tethered to the ends, few principles of base sequence have been yet discovered as determinants of flexibility, or, aside from A tracts, of static bends. DNA curvature (static bend) and flexibility are important issues in understanding indirect readout of DNA sequences by DNA-binding proteins. Use of FRET in relatively short oligonucleotides, 14–25 bases, should in principle allow the effects of specific short sequences in the center of the oligonucleotide to be discerned. Such fluorescence measurements would complement those of NMR or EPR.[31] Assume that one has a straight and rigid DNA of length L. Such a DNA might be one that has internal compensation for overall bending,[32] or be rich in A and T for

[21] J. Wu, K. M. Parkhurst, R. M. Powell, M. Brenowitz, and L. J. Parkhurst, *J. Biol. Chem.* **276,** 14614 (2001).

[22] H. M. Wu and D. M. Crothers, *Nature* **308,** 509 (1984).

[23] H. S. Koo, H. M. Wu, and D. M. Crothers, *Nature* **320,** 501 (1986).

[24] J. C. Marini, S. D. Levene, D. M. Crothers, and P. T. Englund, *Proc. Natl. Acad. Sci. USA* **79,** 7664 (1982).

[25] D. M. Crothers and J. Drak, *Methods Enzymol.* **212,** 46 (1992).

[26] P. J. Hagerman, *Annu. Rev. Biochem.* **59,** 755 (1990).

[27] D. M. Crothers, T. E. Haran, and J. G. Nadeau, *J. Biol. Chem.* **265,** 7093 (1990).

[28] M. A. el Hassan and C. R. Calladine, *J. Mol. Biol.* **259,** 95 (1996).

[29] I. Brukner, V. Jurukovski, M. Konstantinovic, and A. Savic, *Nucleic Acids Res.* **19,** 3549 (1991).

[30] M. Roychoudhury, A. Sitlani, J. Lapham, and D. M. Crothers, *Proc. Natl. Acad. Sci. USA* **97,** 13608 (2000).

[31] T. Okonogi, S. C. Alley, E. A. Harwood, P. B. Hopkins, and B. H. Robinson, *Proc. Natl. Acad. Sci. USA* **99,** 4156 (2002).

[32] J. Bednar, P. Furrer, V. Katritch, A. Z. Stasiak, J. Dubochet, and A. Stasiak, *J. Mol. Biol.* **254,** 579 (1995).

binding of netropsin.[33,34] FRET studies by DoDFRET and ADFRET with various lengths and hence phasings should allow the interdye distributions to be parameterized for extracting the DNA bending alone. In our own work to date, a minimum value of about 4 Å has been found for σ in studying various oligonucleotides, which would set an upper limit on tether motion in 14-mers. Consider a simple model for DNA bending, useful because of its simplicity and the potential for ordering various sequences with regard to flexibility and static bends.

If one considers a rigid rod with a single hinge in the center, the interdye distance will be R, and the contour length L will be considered to include the distance from the ends to the mean position of the attached dye, so R varies from L to some minimum value at the maximum bend angle θ_{max}. We assume the bending follows Hooke's law, so the potential is quadratic in the bend angle. The $P(R)$ for the bending motion alone is a concave function strongly peaked at R corresponding to bending angles of 0 and θ_{max}, because the distance R changes slowly in time as the bend angle approaches 0, and because the dwell time is longest at the turning point, θ_{max}. This is a general feature of any Hooke's law bending model. Superimposed on this $P(R)$ are Gaussians representing the tether motion, weighted by the amplitudes of the bending $P(R)$. The overall result is generally a Gaussian broadened by both the tether motion and by the amplitude of the bending and hence the width of the $P(R)$ for bending. Knowing the tether contribution allows the bending distribution $P(R)$ to be determined. Returning to consideration of the simple central hinge model, one finds solutions in terms of Bessel functions. If one adds to it a static bend, series of Bessel functions characterize the $P(R)$ parameters, although the convergence is relatively rapid. Similarly, a smooth bend model for DNA also involves series of Bessel functions. As an example of how bending information can be extracted from $P(R)$, we examine only the simple model here. For the central hinge model, with no static bend, the $\overline{R}/L$ ratio is given by

$$\overline{R}/L = J_0(\theta_{max}/2) \tag{22}$$

and the reduced $(\sigma/L)^2$ (for only the bend) is given by

$$(\sigma/L)^2 = 0.5 + [0.5J_0(\theta_{max})] - [J_0(\theta_{max}/2)]^2 \tag{23}$$

[33] D. Rentzeperis and L. A. Marky, *J. Am. Chem. Soc.* **115,** 1645 (1993).
[34] D. Rentzeperis, L. A. Marky, T. J. Dwyer, B. H. Geierstanger, J. G. Pelton, and D. E. Wemmer, *Biochemistry* **34,** 2937 (1995).

Significant flexing of the DNA will lead to decreased values of $\bar{R}/L$. Expansion of the Bessel functions leads to excellent approximations for $\bar{R}/L$ and σ/L in terms of θ_{max} (in radians). For $\bar{R}/L \approx 1 - \theta_{max}^2/16$, the error is less than 0.2% out to $\theta_{max} = 60°$; for $\sigma/L \approx 2^{1/2}\theta_{max}^2/16$ the error is <2.4% out to $\theta_{max} = 60°$. These relations show that a plot of $\bar{R}/L$ versus σ/L will be a straight line: $\bar{R}/L = 1 - 2^{1/2}(\sigma/L)$, for a given L, both parameters $\bar{R}$ or σ varying with the square of the maximum bending angle, meaning that either $\bar{R}$ or σ (with a knowledge of L) suffices to determine θ_{max}. Although the parameter equations are more complex for curvature represented by a static bend angle Φ superimposed on the flexibility θ_{max}, an excellent approximation is the straight line:

$$\bar{R}/L = (1 - \Phi^2/8) - [2^{1/2}/\exp(-0.3\Phi)](\sigma/L) \tag{24}$$

Once $\bar{R}$ and σ have been determined, Φ is known, and then θ_{max} can be calculated. The equations linking θ_{max} and σ/L are complex, varying such that as Φ, the static bending increases, σ/L varies with a lower than second power of θ_{max}, becoming nearly linear, for instance, for $\Phi = 90°$. [At $\Phi = 90°$, an excellent approximation is $\theta_{max} = 6.67(\sigma/L)^{0.933}$.] The consequence is that a given value of σ/L corresponds to a much larger angle θ_{max} for $\Phi = 0$ than for increased values of Φ. For this model, not all points with coordinates $(\sigma/L, \bar{R}/L)$ give allowed values of θ_{max} and Φ, but there is a one-to-one mapping in the allowed region, so θ_{max} and Φ are uniquely determined from σ and $\bar{R}$. The details of this simple model are not so important as the fact that FRET-derived $P(R)$ parameters should allow one to obtain values that characterize both static and dynamic bending when used with more realistic bending models,[35–38] although parameterization of the tether motion and knowledge of L are required. Determinations of $P(R)$ as a function of temperature would then be interpretable in terms of the Boltzmann population of energy levels for the bending motion. As in an EPR study of oligonucleotide flexibility,[31] one would expect, from simple statistical thermodynamic considerations, that the FRET parameter σ would be correlated with increased entropy.

[35] P. Wu, B. S. Fujimoto, and J. M. Schurr, *Biopolymers* **26**, 1463 (1987).

[36] L. Song and J. M. Schurr, *Biopolymers* **30**, 229 (1990).

[37] T. M. Okonogi, A. W. Reese, S. C. Alley, P. B. Hopkins, and B. H. Robinson, *Biophys. J.* **77**, 3256 (1999).

[38] T. M. Okonogi, S. C. Alley, A. W. Reese, P. B. Hopkins, and B. H. Robinson, *Biophys. J.* **78**, 2560 (2000).

DNA Bending by Proteins

GCN4 Peptides and AP1 Site Binding

The binding of modified GCN4 peptides (71 amino acids per monomer) to a 14-mer double-stranded DNA oligonucleotide (top strand, TAMRA-5′-GGCTGACTCATTGG-3′-fluorescein, AP1 sequence in boldface) was studied by FRET and by gel electrophoretic phasing techniques[2] to assess the effects of charge modifications in the peptides on the bending of the DNA. The AP1 site is a target for the binding of GCN4, a member of the bZip family of transcription factors, and the binding and bending of the DNA were studied with a view toward correlating the bending with charge variations in the peptides. Thus, residues 17–19 only in the peptides were PAA (neutral), KKK (six positive charges per dimer), and EEE (six negative charges per dimer). Both charge variants showed a greater bend than did the neutral peptide, but σ for the EEE variant was a full angstrom larger than for PAA or KKK. The function $G(t)$ described above was introduced and the area under the curve, μ_G°, provided graphic evidence of the small differences in bend angles. This study was part of ongoing studies into the origins and ultimately the thermodynamics of induced bending arising from charge interactions between the DNA and proteins or protein mimics.[39–42]

TATA-Binding Protein

Strong Promoters. X-ray crystallography has shown that TATA-binding protein (TBP) from Arabidopsis[43] and truncated TBPs from yeast[44] and human[45] bind to the minor groove of strong promoters [such as adenovirus major late promoter (AdMLP), TATAAAAG, and E4, TATATATA] inducing two kinks in the DNA by insertion of pairs of phenylalanines between bases 1 and 2 and bases 7 and 8, resulting in an overall DNA bend of about 80°. DoDFRET was used to characterize $\bar{R}$ and σ for 14-mers before and after binding. The $F_{D(A)}(t)$ decay curve for the complexes showed a marked reduction in $\bar{R}^{16,21}$ and analysis of the distance change

[39] L. J. Maher, Curr. Opin. Chem. Biol. 2, 688 (1998).
[40] L. D. Williams and L. J. Maher, Annu. Rev. Biophys. Biomol. Struct. 29, 497 (2000).
[41] J. K. Strauss-Soukup and L. J. Maher, Biochemistry 36, 10026 (1997).
[42] J. K. Strauss-Soukup and L. J. Maher, Biochemistry 36, 1060 (1998).
[43] J. L. Kim, D. B. Nikolov, and S. K. Burley, Nature 365, 520 (1993).
[44] Y. Kim, J. H. Geiger, S. Hahn, and P. B. Sigler, Nature 365, 512 (1993).
[45] Z. S. Juo, T. K. Chiu, P. M. Leiberman, I. B. Baikalov, A. J. Berk, and R. E. Dickerson, J. Mol. Biol. 261, 239 (1996).

($\Delta \overline{R}$) in terms of a simple two-kink model gave a bend angle in excellent agreement with the crystallographic structure. Time- resolved FRET measurements showed, however, that the E4 interdye distance in the TBP complex was nearly 1 Å less than for AdMLP, with a corresponding bend angle of 71.8° versus 76.2°.[21,46] These differences correspond to overall differences in van't Hoff enthalpies of 16.9 ± 2.2 and 25.1 ± 1.7 kcal/mol for AdMLP and E4, respectively. Combined equilibrium and kinetic studies employing FRET, described below, showed, however, that the bound complex was an equilibrium of at least three bound forms of TBP–DNA.[17,46] Overall, van't Hoff enthalpies thus include enthalpies weighted by the mole fractions of each conformer, which in turn give rise to small heat capacity changes on binding.[47] Kinetic and equilibrium measurements as a function of temperature allowed the thermodynamics of binding and bending to be dissected along the reaction path and interpreted in terms of changes in the bending.

Distance Parameters for Single Base Variants. Several single-base variants of the canonical AdMLP TATA box were studied (A3, T6, C7, G6, and T5, in order of increasing bend angle in solution), which, except for A3, had also been studied by X-ray crystallography.[48] The A3 variant (TAAAAAAG), with an A6 tract, could not be crystallized. The crystallographic studies found little difference in the bending in the solid state—all variants were bent similarly to AdMLP, even C7, which had a Hoogsteen base pair. In contrast, DoDFRET showed marked differences in $\overline{R}$ for the variants when bound to TBP, and a range of bend angles, according to the two-kink model, that extended from about 33 to 80° for AdMLP.[21] The precision of $\overline{R}$ in these studies was 0.1 Å; that of σ, 0.2–0.4 Å. (As a point of reference, for $\overline{R} \approx$ Ro, a precision of 0.3 Å in $\overline{R}$ corresponds to about 1% precision in steady state intensities.[3]) A simple two-state model was proposed to account for the range of structures, in which a given "structure" was considered to derive from an equilibrium between A3 (transcriptionally inactive) and AdMLP (a strong promoter). It was found that a single adjustable parameter allowed satisfactory fits of distance parameters for each variant in terms of the $P(R)$ parameters for A3 and AdMLP. In this description, the structure for a variant is considered to derive from an equilibrium between two forms, as does the bend angle. The average bend angle was found to correlate well with the transcriptional

[46] R. M. Powell, K. M. Parkhurst, M. Brenowitz, and L. J. Parkhurst, *J. Biol. Chem.* **276,** 29782 (2001).

[47] R. M. Powell, K. M. Parkhurst, and L. J. Parkhurst, *J. Biol. Chem.* **277,** 7776 (2002).

[48] G. A. Patikoglou, J. L. Kim, L. Sun, S. H. Yang, T. Kodadek, and S. K. Burley, *Genes Dev.* **13,** 3217 (1999).

activity of the various promoters,[21] and it was proposed that consideration of the binary interaction of TBP and promoter was crucial for understanding transcriptional activity. Furthermore, the equilibrium description of transcriptional activity and the $\bar{R}$ values for the various promoters led to the proposal that the variation of activity with sequence was related to structural constraints placed on the addition of subsequent initiation factors by these binary TBP–DNA structures.

Effects of Osmolytes on Distance Parameters and Energetics of DNA–TBP Bending. Variants T6 and C7 were studied together with AdMLP[49] in equilibrium binding studies with yeast TBP as a function of osmolyte (glycerol, ethylene glycol) in order to probe water release associated with the binding.[50–54] The overall binding of TBP and DNA is endothermic and is driven by large entropy changes for temperatures less than 30°. These entropy changes presumably derive in part from water and salt release from the minor groove as well as changes in salt and water binding in the compressed major groove. The distance parameters and the equilibrium constant were only slightly changed for AdMLP. On the other hand, addition of osmolyte resulted in large progressive decreases in $\bar{R}$ (increases in the bend angle) on binding of the variants to TBP as well as increases in binding affinity. Indeed, the bend angles approached closely that of AdMLP at 3 M osmolyte. Again, the progressive sequence of structures could be fit satisfactorily in the same manner as for the single-base variants as an equilibrium between A3 and AdMLP-like conformations.[49] This two-state model showed that only 0.06 kcal/mol separated the two conformations for the T6 variant at 1 M osmolyte, and because crystals of TBP and DNA are grown from solutions of osmolytes, an explanation for the differences between solution and crystal structures was proposed. The structure of A3 in solution, on the other hand, was not influenced by osmolyte, consistent with the lack of success in crystallizing a TBP complex of this variant.

Kinetics and Energetics of DNA Binding and Bending by TBP. DoD-FRET and steady state FRET measurements of binding equilibria and kinetics have provided a map of the structural and energetic changes for the sequential binding and bending of AdMLP, E4, and C7 by full-length yeast TBP. DoDFRET provided $P(R)$ parameters for the double-stranded DNA

[49] J. Wu, K. M. Parkhurst, R. M. Powell, and L. J. Parkhurst, *J. Biol. Chem.* **276,** 14623 (2001).
[50] S. N. Timasheff, *Annu. Rev. Biophys. Biomol. Struct.* **22,** 67 (1993).
[51] S. N. Timasheff, *Proc. Natl. Acad. Sci. USA* **95,** 7363 (1998).
[52] S. N. Timasheff, *Adv. Protein Chem.* **51,** 355 (1998).
[53] V. A. Parsegian, R. P. Rand, and D. C. Rau, *Methods Enzymol.* **259,** 43 (1995).
[54] V. A. Parsegian, R. P. Rand, and D. C. Rau, *Proc. Natl. Acad. Sci. USA* **97,** 3987 (2000).

(dsDNA) sequences before and after binding by TBP. For AdMLP and E4, there were no changes in steady state FRET intensities with temperature. On the other hand, the steady state intensities for the C7–TBP complex were sensitive to changes in temperature, consistent with an increase in the bend angle as the temperature increased.[47] Equilibrium constants for TBP binding were determined as a function of temperature over the range $10-30°$ and the data were well fit by a simple linear van't Hoff plot. In all cases the binding was strongly endothermic (17 to 25 kcal/mol) and driven overall by an increase in entropy.

The linear mechanism found to fit the combined equilibrium and kinetic data for AdMLP and for E4 was[46]

$$TBP + DNA \leftrightarrow I_1 \leftrightarrow I_2 \leftrightarrow final \tag{25}$$

The species I_1 is not to be confused with an encounter complex; it is a complex present in detectable amounts at equilibrium. In the three-step four-species scheme, the kinetics of binding, combined with fitted relative quantum yields, showed that the first step involved simultaneous binding and bending of the DNA to produce the first detectable complex, I_1. Although they shared a common mechanism, the individual steps in the reaction had quite different thermodynamic changes for AdMLP and E4. Figure 7 shows enthalpy and entropy changes for AdMLP and E4 along the reaction coordinate. Base differences between these two strong promoters residing in the latter half of the TATA box were used to account for the thermodynamic differences. Recall that DoDFRET showed that there was overall a significantly decreased bend angle for E4 and that the overall enthalpy change was larger for E4. For C7, a three-step mechanism was employed, but with a nonproductive equilibrium between TBP plus DNA leading to a shallowly bent complex, and with one intermediate between TBP plus DNA and the final complex (see Fig. 8).[47] Kinetic analyses revealed major changes, especially for C7, of the relative populations of DNA–TBP conformers as a function of temperature.

The designation "final" in Eq. (25) is meant only to indicate that it is the terminal species in a linear scheme and does not imply, for instance, correspondence to the crystallized conformation for AdMLP. In some instances, it is, as for E4,[46] not the thermodynamically most favored conformer at all temperatures. The conformer I_2 for AdMLP and E4 is never present in amounts much greater than 1–2% mole fraction, but it is absolutely required for adequate fits to the kinetic data. On the other hand, the intermediate I_1, for both AdMLP and E4, can represent 0.8 mole fraction of the bound DNA for about 1 min and we have proposed that this species, with bent DNA, is the biologically relevant conformer for subsequent assembly of the preinitiation transcription complex.[17] What is

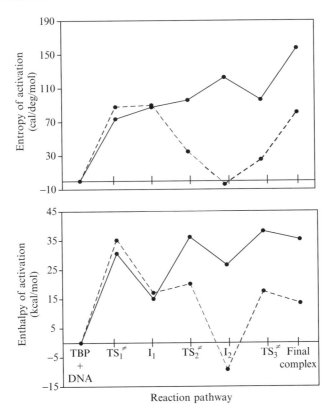

FIG. 7. Thermodynamic profiles for TBP plus DNA reactions of E4 (——) and AdMLP (- - - -) at 25°, 1 M standard states. Units are kcal/mol for $\Delta H°$ and cal/mol·K for $\Delta S°$. The species I_1, I_2, and Final Complex, pertaining to Eq. (25), are indicated, as are the transition states.

unknown at the present time are the precise $P(R)$ parameters for this intermediate. These should be accessible, however, using rapid mixing to prepare I_1 at high mole fraction and carrying out multiple measurements of trFRET, using a rapidly pulsed diode laser and time-correlated single-photon counting. In the kinetic fits, the relative quantum yields of the three bound complexes with respect to the initial DNA were treated as fitting parameters in the global fitting scheme. For E4 it appears that, to the extent that the steady state fitted quantum yield reflects $\overline{R}$, I_1 is slightly less bent than the final conformer, and from Fig. 7 it can be seen that a further increase in enthalpy separates I_1 and the final complex. Figure 8 shows that a single substitution in the TATA box to give the C7 promoter, results in a markedly different thermodynamic reaction profile.

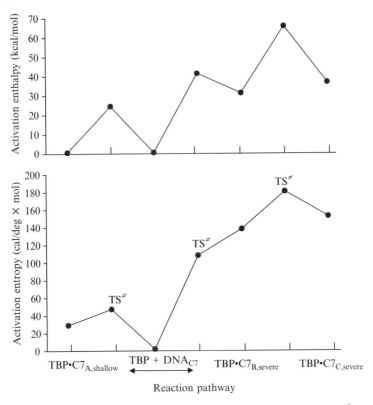

FIG. 8. Thermodynamic profiles for TBP plus DNA for the C7 variant, $25°$ and $1\ M$ standard state. Units are kcal/mol for $\Delta H°$ and cal/mol·K for $\Delta S°$. Species and transition states are shown.

Kinetic data, on which much of the detailed energetics were based, were collected in the following way. Association kinetics were determined by fluorescence stopped flow. Dissociation kinetics were carried out by rapid manual mixing in a steady state spectrofluorimeter (PTI Alpha Scan) in which a large excess of unlabeled dsDNA was mixed with equilibrated DNA–TBP complexes. Both forward and reverse reactions were studied as a function of TBP and for the dissociation process, as a function of unlabeled DNA. The dissociation reaction was found to be complex, consisting of both replacement and displacement processes. These processes had to be separated in order to obtain correct fits and interpretation for the above-described model that allows only replacement.

The detailed fitting of the kinetics that provided the energetics of binding and bending along the reaction coordinate proceeded in the following way. Decomposing the kinetic curves into eigenvalues and associated amplitudes and fitting these quantities was not satisfactory. Rather, the kinetic curves were fitted. No constraints were placed on the temperature dependence of the equilibrium data; rather, the equilibrium curves at each temperature were fitted together with the kinetics. The rate constants, on the other hand, were used to describe the equilibria, and were thus constrained by the requirement to fit both kinetics and equilibria. The rate constants were written in terms of a rate constant at 30° and an enthalpy of activation to link values of a given rate constant as a function of temperature. The resulting constraints in the global fitting are discussed elsewhere.[17,46] Because the kinetic scheme is a relatively simple first-order network, it can be solved exactly by diagonalizing the K matrix (which requires special treatment because it is not symmetric),[55,56] by an exponential matrix expansion,[57,58] or by use of theorems from matrix algebra commonly used in linear systems analysis. We have employed the latter procedure for TBP–DNA kinetics[17] and previously in studies of hemoglobin ligand kinetics.[59,60] Because determination of the detailed energetics requires a rapid integration method, we provide here a brief but sufficient description. We require only traces of powers of the $n \times n K$ matrix to establish the coefficients of the characteristic polynomial, the roots of which are the eigenvalues, a procedure originally due to Bôcher.[61] Knowing the initial concentration vector, $A(0)$, $A(t) = E \cdot A(0)$, and from $A(t)$ the measured scalar response $R(t)$ can be obtained in terms of intrinsic variables (quantum yields, molar absorptivities, etc.) of the species in $A(t)$. E in turn can be written as a sum of matrices, $1, 2, \ldots, n$, where n is the number of eigenvalues, including the zeroth or equilibrium eigenvalue, where $E = E_1 + E_2 + \cdots + E_n$. If λ_i denotes the ith eigenvalue of the $-K$ matrix, then we can write $E_i = \exp(\lambda_i t) E_i'$. There are several ways of developing the E' matrices, or their equivalent, including residue theory, partial fractions,[62] or Sylvester's theorem,[63] all of which must give the same result, although the matrix expressions can appear different. Sylvester's approach

[55] P. C. Jordan, "Chemical Kinetics and Transport," pp. 153–157. Plenum, New York, 1979.
[56] F. A. Matsen and J. L. Franklin, *J. Am. Chem. Soc.* **72,** 3337 (1950).
[57] T. M. Zamis, L. J. Parkhurst, and G. A. Gallup, *Comput. Chem.* **13,** 165 (1989).
[58] I. B. C. Matheson, L. J. Parkhurst, and R. J. De Sa, *Methods Enzymol.* (accepted 2003).
[59] K. M. Parkhurst, Ph.D. dissertation. University of Nebraska, 1991.
[60] Y. Gu, Ph.D. dissertation. University of Nebraska, 1995.
[61] M. Bôcher, "Introduction to Higher Algebra." Macmillan, New York, 1931.
[62] Research and Education Association Staff, *in* "Theory of Linear Systems" (R. Fratila, ed.), pp. 46–121. Research and Education Association, New York, 1982.
[63] R. A. Frazer, W. J. Duncan, and A. R. Collar, *in* "Elementary Matrices," pp. 78–87. Cambridge University Press, Cambridge, 1938.

is notationally the simplest: E_i' can be written in terms of a numerator and a denominator. The denominator is a product of $n - 1$ successive differences of the various eigenvalues with respect to λ_i: $\prod(\lambda_i - \lambda_j)$, $i \neq j$. The numerator is a product of $n - 1$ matrices, written in terms of the **K** matrix and modified by eigenvalues other than λ_i written on the principal diagonal. Thus with **1** as the identity matrix, the numerator becomes $\prod(\mathbf{K} + \lambda_j\mathbf{1})$, where the product is over $n - 1$ values, with $j \neq i$. In the actual data fitting, the matrix routine is a subroutine in a Simplex program, and the χ^2 minimized in the Simplex is over all temperatures, TBP concentrations, and both kinetic and equilibrium data. The simplex variables correspond to the high-temperature rate constants and the associated activation enthalpies and the relative quantum yields of the species. Errors are determined by fitting extensive simulations of the data[64] with Gaussian distributed random noise added to curves that represent satisfactory fits to the data according to F tests.[65,66]

Summary

Parameters that characterize a distance distribution $P(R)$ can be obtained from time-resolved FRET measurements. These measurements can involve various combinations of donor-detected FRET and acceptor-detected FRET constrained by steady state emission intensity differences between the donor and that of the donor in the presence of an acceptor. Highly precise average interdye distances $\bar{R}$ can ultimately lead to precise intramolecular distances in solution. The width of the $P(R)$ distribution, σ, preferably and more precisely after removal of the tether contributions, yields a measure of conformational equilibria and of conformational dynamics of the macromolecule to which the probes are attached. FRET measurements combined with equilibrium determinations and with rapid-mixing or relaxation kinetics provide structure-energy, entropy profiles of intermediates and transition states along the reaction coordinate.

Acknowledgments

I thank Mr. Roberto Fabio Delgadillo for Fig. 6, and acknowledge support from NIH Grants GM 59346, CA 76049, and RR 15635.

[64] W. H. Press, B. P. Flannery, S. A. Teukolsky, and W. T. Vetterling, *in* "Numerical Recipes," pp. 503, 529–538. Cambridge University Press, New York, 1989.
[65] N. R. Draper and H. Smith, *in* "Applied Regression Analysis," pp. 63–69, 280–290. John Wiley & Sons, New York, 1966.
[66] Y. Bard, *in* "Nonlinear Parameter Estimation," pp. 170–192. Academic Press, New York, 1974.

Author Index

Numbers in parentheses are footnote reference numbers and indicate that an author's work is referred to although the name is not cited in the text.

A

Abbasi, A., 86, 98(19)
Abbott, J., 214, 220(19)
Abelson, J., 214, 217(18), 220(18)
Abeygunawardana, C., 156, 157, 159(24)
Abola, E. E., 207
Acharya, A. S., 42
Ackers, G. K., 3, 4(3), 5, 6(1; 2), 7, 9, 9(5), 10(4), 12(4), 13, 14(1; 12; 13), 15, 15(12), 17, 18(20), 19, 19(13), 21(27), 22(13; 27), 23(27), 24(7), 25, 25(14), 26(4; 19; 23), 29, 41, 51, 53(72), 74, 75, 75(20), 169, 175, 220, 220(31), 231(32)
Adhya, S., 178
Agrawal, B. B. L., 110
Alberti, S., 197
Alberty, R. A., 238, 239(6)
Ali, S. A., 86, 98(19)
Alley, S. C., 252, 254, 254(31)
Almo, S. C., 49
Amiconi, G., 59
Andersen, M. E., 55, 56, 56(2)
Anderson, C. F., 146
Andreu, J. M., 101
Antonini, E., 28(1), 29, 55, 57
Appella, E., 185
Arnarp, J., 119
Arnone, A., 31, 32(15), 33(16), 41, 54(15; 16)
Arrington, C. B., 152(4), 153, 154(4)
Asakura, T., 47
Ascenzi, P., 62
Ascoli, F., 55, 57
Atha, D. H., 74

B

Bacon, D. J., 30, 44(5)
Bai, Y., 152(3), 153, 154(3)
Baichoo, N., 215
Baikalov, I. B., 255
Baker, C. H., 138
Baker, E. N., 73, 76(18)
Baker, H. M., 73, 76(18)
Baldwin, J., 31, 49(12), 169, 172(36)
Bandyopadhyay, S., 25, 186
Banerjee, R., 178
Banik, U., 186
Bard, Y., 262
Barker, D. F., 211, 218(9)
Barrick, D., 28, 31, 43, 43(13; 14), 45(13; 14), 46(13; 14)
Barros, A. C. H., 110
Barry, J. K., 199, 200, 200(21), 201(23), 204, 205(21)
Barshop, B. A., 233
Baskerville, M., 183
Baudras, A., 141
Bax, A., 37, 38, 39(26), 40, 51(32), 52(32), 53(32), 156, 157, 159(23)
Beckerbauer, L., 182
Beckett, D., 25, 209, 213, 214, 214(17), 216, 217(26), 219, 219(15), 220(19; 30; 31), 221(31), 222, 223(15; 36), 224(27), 226(27), 227(27), 228, 228(20; 27), 229, 231(38), 232(38)
Beckman, E., 99, 104(46)
Bednar, J., 252
Bell, C. E., 198, 205
Bell, D. A., Jr., 41, 42(38)
Beltramini, M., 86, 98, 98(20)
Benazzi, L., 12, 15
Bencomo, V. V., 107, 110(3)
Benesch, R. E., 69

263

Subject Index

A

Allosterism
  biotin repressor, *see* Biotin repressor
  lactose repressor studies
    functional intermediate detection
      indications, 190, 194
      overview, 188–190
      thermodynamic description of end
        points and binding equilibria,
        190, 192–193
    intermediate processes in DNA
      binding mechanisms, 196
    molecular dynamics simulations,
      206–208
    prospects for study, 208–209
    site-direced mutagenesis studies
      communication disruption by adding
        glycines, 204
      communication disruption by
        clamping, 204–205
      design of mutants, 201
      random mutagenesis, 202–203, 206
      testing interactions critical to
        allosteric response, 200
      tryptophan scanning, 201, 204
    subunit contributions to binding and
      allostery, 196–197, 199

B

Biotin repressor
  biotin binding
    affinity, 213
    mechanism, 214
  conformational probing of ligand
    binding effects using hydroxyl
    radical cleavage
    cleavage reactions, 217
    data analysis, 218
    gel electrophoresis of cleavage
      products, 217
    overview, 215–216

    protein preparation and
      phosphorous-32 labeling,
      216–217
  correlation of dimerization and
    DNA binding
    DNase I footprinting, 224
    kinetic analysis using time-resolved
      DNase I footprinting
      footprinting, 229–231
      modeling, 228–229, 233–234
      progress curve generation and
        analysis, 231–233
    sedimentation equilibrium, 225, 227
    thermodynamic linkage, 224, 227
  functions, 211
  operator binding, 214
  site-directed mutagenesis studies
    correlation of loss of assembly and
      DNA-binding affinity
      reduction, 222
    DNase I footprinting, 220
    mutant protein preparation, 219–220
    sedimentation equilibrium analysis of
      assembly, 220–221
  small molecule effectors in transcription
    regulation, 209–211
  structure, 213, 222–223

C

Circular dichroism, cyclic AMP receptor
  protein cooperative binding studies,
  144–145
Concanavalin A, ligand
  specificity, 107
Cooperativity
  hemocyanin, *see* Hemocyanin
  hemoglobin, *see* Hemoglobin
  multiple binding of ligands to linear
    biopolymer, two-ligand
    multiple-binding model
    binding isotherms, 151
    elementary units

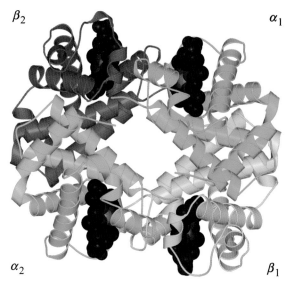

β_2 α_1

α_2 β_1

BARRICK *ET AL.*, CHAPTER 2, FIG. 1. The crystal structure of Hb A. α subunits are shown in green and turquoise, and the β subunits are shown in yellow and orange. The $\alpha_1\beta_1$ dimer is on the right; the $\alpha_2\beta_2$ dimer is on the left. The four heme groups are shown as van der Waals spheres. Coordinates are from pdb file 1A3N. This image was prepared using MOLSCRIPT and Raster3D.

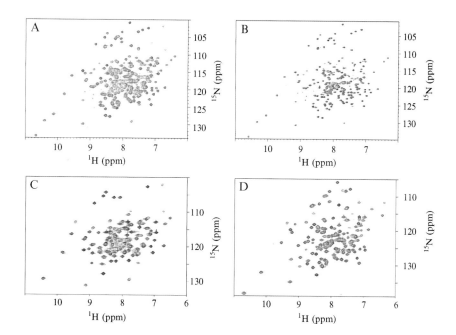

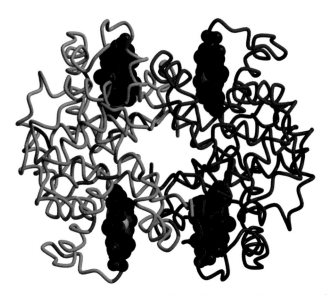

BARRICK *ET AL.*, CHAPTER 2, FIG. 4. A comparison of the extent of resonance assignments in HbCO A using ^{1}H NMR *(left)* and by more recent multinuclear [^{13}C,^{15}N] experiments *(right)*. Dark shading (blue for α subunits, red for β subunits) indicates residues for which at least one resonance assignment has been made; light shading indicates residues that are unassigned. The $\alpha_1\beta_1$ dimer on the left shows assignments made before the application of multinuclear multidimensional NMR to isotopically labeled HbCO A, whereas the $\alpha_2\beta_2$ dimer on the right includes assignments made as a result of these methods (Lukin *et al.* backbone resonances have been deposited at BioMagResBank). Assignment information is superposed on the crystal structure of HbCO A in the R2 form (1BBB). This image was prepared using MOLSCRIPT and Raster3D.

BARRICK *ET AL.*, CHAPTER 2, FIG. 2. [^{15}N, ^{1}H] single-bond correlation spectroscopy of uniformly labeled and chain-selective HbCO A. (A) [^{15}N, ^{1}H]-HSQC spectrum of uniformly ^{15}N, ^{2}H-labeled rHbCO A at 600 MHz in 95% H$_2$O–5% D$_2$O. ^{1}H signals result from protons that can exchange with H$_2$O, that is, peptide and side-chain (N)H groups. Although resolution is greatly improved over the analogous 1D ^{1}H NMR spectrum, considerable overlap remains. (B) TROSY spectrum of uniformly ^{15}N, ^{2}H-labeled rHbCO A at 600 MHz, which results in a significant decrease in linewidth, thus decreasing resonance overlap compared with the analogous HSQC spectrum. (C and D) [^{15}N,^{1}H]-HSQC spectra of ^{15}N, ^{2}H, ^{13}C-labeled rHbCO A samples in which labeling is restricted to the α subunits (blue) or the β subunits (red) respectively, thus decreasing overlap compared with the spectrum of the uniformly ^{15}N-labeled sample. (C) and (D) were reproduced from Fig. 2 of Simplaceanu *et al.*, with permission.

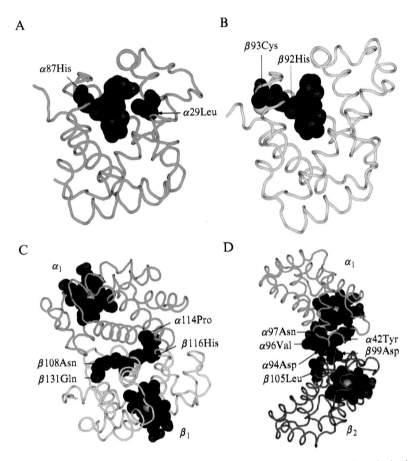

BARRICK *ET AL.*, CHAPTER 2, FIG. 5. Locations of several site-specific substitutions engineered into recombinant Hb A. (A and B) Residues substituted in the heme pockets of the α subunit (yellow) and β subunit (green), respectively, with heme groups represented by van der Waals spheres. (C) Residue substitutions at the $\alpha_1\beta_1$ interface (α and β subunits in green and yellow, respectively). Residue substitutions at the $\alpha_1\beta_2$ interfaces (α and β subunits in green and orange, respectively). This image was prepared using MOLSCRIPT and Raster3D.

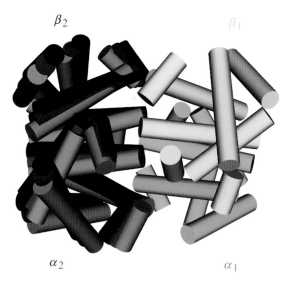

β_2 β_1

α_2 α_1

BARRICK *ET AL.*, CHAPTER 2, FIG. 6. The quaternary structure of HbCO A in solution, compared with the R and R2 crystal structures. The $\alpha_1\beta_1$ dimers (green and yellow, *right*) of each of the three structures were superposed, facilitating comparison of the relative orientations of the $\alpha_2\beta_2$ dimers (red and blue, *left*). For the high-salt R-state crystal structure 1IRD, the $\alpha_2\beta_2$ dimer is shown with dark shading; for the low-salt R2 crystal structure (1BBB), the $\alpha_2\beta_2$ dimer is shown with light shading; and for the solution structure determined using residual dipolar couplings, the $\alpha_2\beta_2$ dimer is shown with intermediate shading. Reproduced from Fig. 2 of Lukin *et al.*, with permission.